Routledge Revivals

The Comparative Economics of Plantation Forestry

Plantation forestry is the planting, managing, and harvesting of trees for the production of industrial wood. Originally published in 1983, the principal focus and contribution of the study lies in Roger Sedjo's examination of the economic returns in twelve forest regions throughout the world. The results of the analysis strongly demonstrate the feasibility of major expansion of plantation forestry in a number of areas around the world and suggest the likelihood of major shifts in the principal supply areas. The results also have potentially important implications for countering deforestation. This title will be of interest for students of Environmental Studies.

The Comparative Economics of Plantation Forestry
A Global Assessment

Roger A. Sedjo

First published in 1983
by Resources for the Future, Inc.

This edition first published in 2016 by Routledge
2 Park Square, Milton Park, Abingdon, Oxon, OX14 4RN
and by Routledge
711 Third Avenue, New York, NY 10017

Routledge is an imprint of the Taylor & Francis Group, an informa business

© 1983, Resources for the Future, Inc.

All rights reserved. No part of this book may be reprinted or reproduced or utilised in any form or by any electronic, mechanical, or other means, now known or hereafter invented, including photocopying and recording, or in any information storage or retrieval system, without permission in writing from the publishers.

Publisher's Note
The publisher has gone to great lengths to ensure the quality of this reprint but points out that some imperfections in the original copies may be apparent.

Disclaimer
The publisher has made every effort to trace copyright holders and welcomes correspondence from those they have been unable to contact.

A Library of Congress record exists under LC control number: 83042906

ISBN 13: 978-1-138-11973-4 (pbk)
ISBN 13: 978-1-138-10149-4 (hbk)
ISBN 13: 978-1-315-65216-0 (ebk)

The Comparative Economics
of Plantation Forestry

Three traditional and nontraditional forest-producing regions of the world selected for plantation forestry assessment.

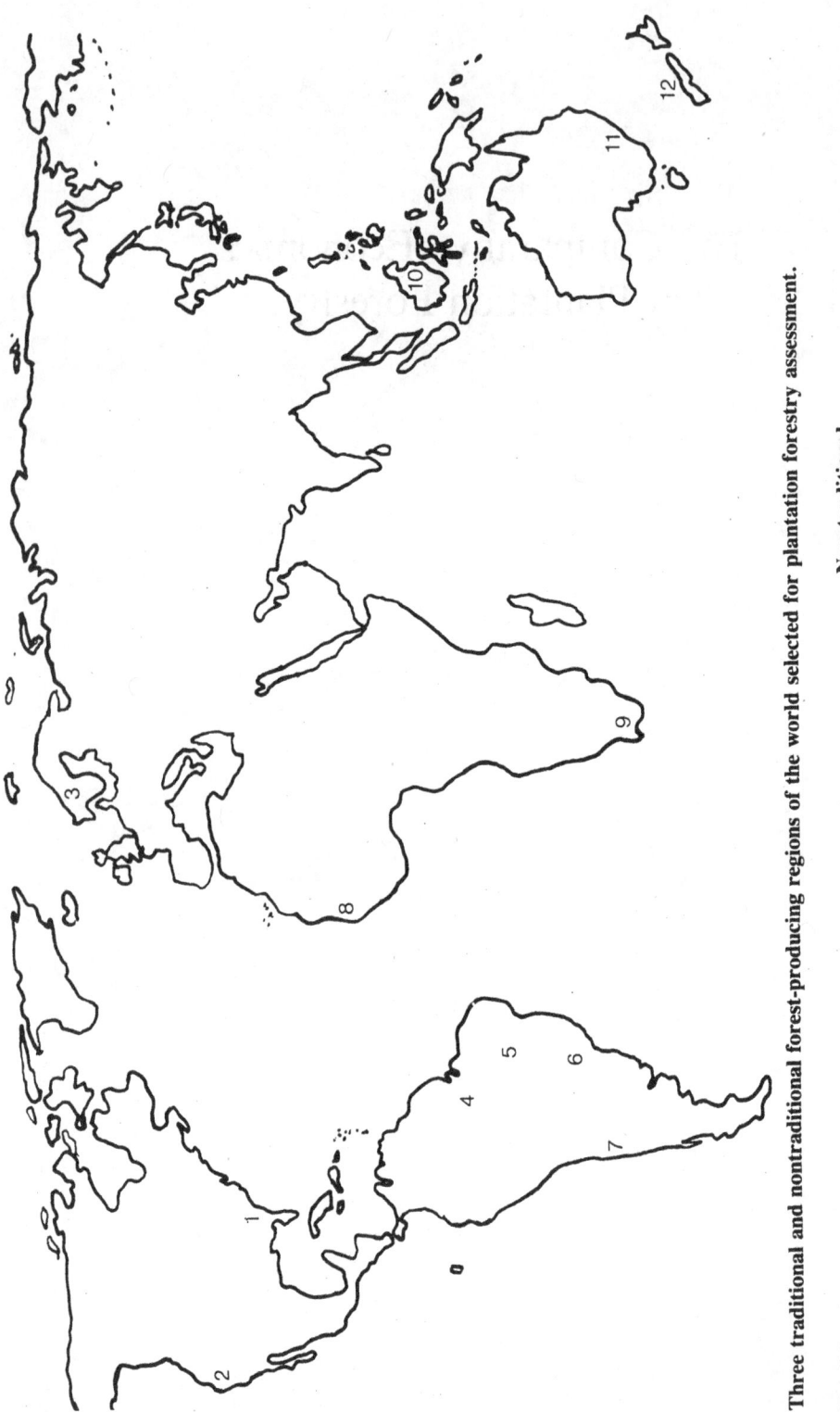

Traditional
1. U.S. South
2. Pacific Northwest
3. Nordic region (Norway, Sweden, Finland)
4. Amazonia
5. Central Brazil
6. Southern Brazil

Non traditional
7. Chile
8. West Africa (Gambia and Senegal)
9. South Africa
10. Borneo
11. Australia
12. New Zealand

The Comparative Economics of Plantation Forestry
A Global Assessment

ROGER A. SEDJO

RESOURCES FOR THE FUTURE / WASHINGTON, D.C.

Library of Congress Cataloging in Publication Data

Sedjo, Roger A.
  The comparative economics of plantation forestry.

  Includes index.
  I. Tree farms—Economic aspects.   2. Comparative
economics.   I. Title.   II. Title: Plantation forestry.
SD393.S43   1983   338.1'749   83–42906
ISBN 0–8018–3107–5

Copyright © 1983 by Resources for the Future Ine

Distributed by The Jonns Hopkins University Press, Baltimore, Maryland 21218

Manufactured in the United States of America

Published August 1983

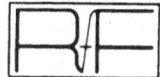 RESOURCES FOR THE FUTURE, INC.
1755 Massachusetts Avenue, N.W., Washington, D.C. 20036

DIRECTORS

M. Gordon Wolman, *Chairman*

Charles E. Bishop
Roberto de O. Campos
Anne P. Carter
Emery N. Castle
William T. Creson
James R. Ellis
Jerry D. Geist
John H. Gibbons
David S. R. Leighton
Franklin A. Lindsay

Vincent E. McKelvey
Richard W. Manderbach
Laurence I. Moss
Mrs. Oscar M. Ruebhausen
Leopoldo Solís
Janez Stanovnik
Carl H. Stoltenberg
Russell E. Train
Robert M. White
Franklin H. Williams

HONORARY DIRECTORS

Horace M. Albright
Edward J. Cleary
Hugh L. Keenleyside

Edward S. Mason
William S. Paley
John W Vanderwilt

OFFICERS

Emery N. Castle, *President*
Edward F. Hand, *Secretary-Treasurer*

Resources for the Future is a nonprofit organization for research and education in the development, conservation, and use of natural resources, including the quality of the environment. It was established in 1952 with the cooperation of the Ford Foundation. Grants for research are accepted from government and private sources only on the condition that RFF shall be solely responsible for the conduct of the research and free to make its results available to the public. Most of the work of Resources for the Future is carried out by its resident staff; part is supported by grants to universities and other nonprofit organizations. Unless otherwise stated, interpretations and conclusions in RFF publications are those of the authors; the organization takes responsibility for the selection of significant subjects for study, the competence of the researchers, and their freedom of inquiry.

This Research Paper is a product of RFF's Renewable Resources Division, Kenneth D. Frederick, director. Research Papers are intended to provide prompt distribution of research having a narrower focus or a greater technical nature than RFF books.

Contents

Foreword   ix
Acknowledgments   xi
1   Introduction   1
2   The Representative Plantation   9
3   Forest Plantation Simulation Model   16
4   Extramodel Factors Affecting Investment in Industrial Forest Plantations   26
5   Model Results: Base Case   36
6   Sensitivity Analysis   48
7   Implications of the Results   80
8   Ecological Implications of Tropical Plantation Forestry, by *Gary S. Hartshorn*   84
Appendixes
    A: Representative Plantation Regimes: Costs and Yields   97
    B: Mathematical Formation of the Model   134
    C: International Transport Cost Methodology   138
    D: Stumpage Prices   147
Index   159

Foreword

Growing demands for wood and its by-products together with the search for new agricultural lands have accelerated cutting or intentional burning of the world's forests. Once cleared of their trees some of these lands are given over to crops, livestock, or more-intensive human uses; they are lost to forestry for the foreseeable future. Other lands may be abandoned once the usable wood has been taken or after a brief period of farming to profit from the fertility of the soil which was built up over many years by the forest cover but is depleted after just a few years of farming. Once left alone, the forest often will regenerate itself. But it may take the better part of a century to restore most of the usable wood to unmanaged lands. And in fragile ecological areas, logging, especially where it is followed by farming, may leave the lands incapable of naturally generating their forest cover and essentially useless for other human needs.

Gradual depletion of what once seemed to be a limitless stock of trees has contributed to the long-term increase in the real price of many forest products. These trends in stocks and prices have led to growing concerns about the cost and adequacy of future wood supplies. Deforestation is also the source of some major environmental concerns. It is blamed for an accelerating reduction in species diversity, for erosion and downstream flooding, for destruction of soil fertility, and for reduction in the globe's capacity to absorb carbon dioxide, the buildup of which could lead to major climatic changes.

There is little prospect that demand will slacken to ease the pressures on the world's forests. There are strong prospects, however, for increasing the yield from forest lands. Higher yields make it possible to satisfy demand from a smaller land base and perhaps to halt the long-term reductions in forest inventories and the rising real cost of wood.

Plantation forestry, defined as the planting, managing, and harvesting of trees for the production of industrial wood, is capable of producing dramatic increases in wood yields. A high-yielding forest plantation can produce 15 to 20 cubic meters of industrial roundwood per year on good forest land. With such yields, which are 30 to 40 times the estimated average yield on the world's forests in 1978, it would be possible to satisfy

FOREWORD

wood needs on just a small fraction of the land now devoted to forests. This potential, however, is not close to being realized. Even though plantation forests are found in many areas of the world, the total acreages involved are small and their contribution to the world's wood supplies also is small.

The future role of plantation forestry both within specific areas of the world and in terms of its overall contribution to global wood supplies will depend in large part on the two issues addressed in this study—the economic feasibility of such plantations in various areas of the world and the long-term ecological impacts of such forestry regimes in tropical areas. The principal focus and contribution of the study lies in Roger Sedjo's examination of the economic returns to representative plantation forests in twelve regions throughout the world. In the final chapter, Gary S. Hartshorn analyzes the ecological implications of tropical plantation forestry.

The results of the analysis strongly demonstrate the feasibility of major expansion of plantation forestry in a number of areas around the world and suggest the likelihood of major shifts in the principal supply areas. The results also have potentially important implications for countering the trend toward deforestation. It is important, however, not to lose sight of the very tentative nature of the results. The methodology used to examine the ability of various regions of the world to compete in the production of sawtimber and wood pulp in three principal markets reflects recent global trade practices in forest products. These practices, and therefore some of the assumptions underlying the stumpage prices, calculated for different regions could be very different in the future. Also underlying the results are the data on the costs of and the biological yields to various forestry practices in different regions of the world. The collection and analysis of these data represent an important contribution of the study. Nevertheless, many of these numbers rest on very limited experience, and there is little attempt to anticipate changes in costs and technologies that undoubtedly will affect the future profitability of investments in plantation forestry. While Roger Sedjo makes an important contribution to understanding the potential of plantation forestry in this volume, an equally important contribution of the study may be the stimulus it provides others to refine the data and analysis initiated here.

International forestry issues have been a major area of research at RFF since the Forest Economics and Policy Program was established in 1977 and Dr. Sedjo joined the staff as director of that program. Other RFF publications stemming from that line of work are *Issues in U.S. International Forest Products Trade* and *Postwar Trends in U.S. Forest Products Trade*. Roger Sedjo's on-going analysis of long-term global timber supply continues this general area of research at RFF. The Forest Service and the Weyerhaeuser Company Foundation are major supporters of RFF's Forest Economics and Policy Program.

March 1983

Kenneth D. Frederick
Director
Renewable Resources
Division

Acknowledgments

Numerous organizations and individuals provided assistance in the conduct of this study. Gary S. Hartshorn of the Tropical Science Center in Costa Rica wrote the chapter on the "Ecological Implications of Tropical Plantation Forestry." The Weyerhaeuser Company was most helpful, with assistance and information provided by Jack Wolff, R. N. Pierson, Gilbert Baker, William Gladstone, George Stabler, Bruce Lippke, and many others. Substantial assistance was provided on a number of international trips associated with the study. John Zivnuska, Hans Gregersen, and G. Robinson Gregory all made helpful suggestions regarding overseas contacts. The following organizations and individuals provided useful assistance and information: in New Zealand, John Tustin, Thomas Fraser, and W. R. J. Sutton of the Forest Research Insitute, and Brian Allison of New Zealand Forest Products Ltd.; in Indonesia, Norman E. Johnson of P. T. International Timber; and in Colombia, William Ladrach and Millan Gutierrez of Carton de Colombia. In Brazil, William Beattie was most helpful in facilitating meetings with appropriate organizations and individuals. Pieter Prange of Olinkraft Celulose e Papel Ltda., together with Mario Colombelli of Duratex and Luiz J. Murat, Jr., of Suzano Feffer, all representing the Associacão Nacionál dos Fabricantes de Papel e Celulose, provided invaluable information and data on a variety of topics related to Brazilian plantations. Mario Ferreira of the Universidade São Paulo and particularly Antonia Natal Goncalves also provided useful assistance. In Chile, assistance was provided by Sergio Silva of the Ministerio de Agricultura, together with Sebastian Hudson and Alberto Lira of the National Forestry Corporation (CONAF). In addition, Andres Tienken of Forestal Celco Ltda. and particularly Bertram Husch of the United Nations Food and Agriculture Organization provided useful information about Chilean forest plantations. Also, thanks are due to Frank Wadsworth and Leon Liegel of the Forest Service Institute of Tropical Forestry in Puerto Rico; Kristi Sutela of the Finnish Forestry Information Service; Michael Hall, A.P.M. Forest Proprietary Ltd. (Australia); P. T. Unwin of the

Forestry Commission of Tasmania; and Richard Jaffie of Robert R. Nathan Associates.

Robert Skillings, Robert Goodland, and Robert Fishwick of the World Bank and David Darr, George Dutrow, Stanley Krugman, Richard Haynes, Thomas Mills, and Les Witmore of the U.S. Forest Service all provided assistance. Mark Goforth of the Forest Service furnished the basic computer model, which was modified only slightly for use in this study, while Robert Curtis of the Forest Service Forest Science Laboratory provided data and a number of computer runs of various Douglas fir regions.

Acknowledgment is also due to those individuals not mentioned previously who commented on all or part of the earlier drafts of this manuscript or provided other special assistance. These include Ake Andersson, University of Umea in Sweden; John Barber, Society of American Foresters; Marion Clawson, Resources for the Future; Perry Hagenstein, Consultant; Thomas Marcin, Forest Products Laboratory; Jack Muench, National Forest Products Association; Ashley Selby, Finnish Forest Research Institute; Harold Wisdom, Virginia Polytechnical Institute; and Bruce Zobel, North Carolina State University; and Sally Skillings of Resources for the Future. Additional thanks are extended to John Zivnuska of the University of California at Berkeley and to Lester Holley of North Carolina State University for their thoughtful and thorough reviews of earlier versions of this manuscript.

Particular thanks must be extended to Samuel Radcliffe, who worked on the early phases of this project. In addition, a special thanks to my research assistant, Catherine Tunis, who joined the project in midstream and was able to quickly master the intricacies of both the computer model and the plantation regimes, and to Donna Stauffer, who also provided assistance.

Furthermore, I am grateful to John Mankin, Lorraine Van Dine, and Cindy Stokes for their efforts typing the manuscript. Dorothy Sawicki deserves a special acknowledgment for her very substantial editing effort.

Finally, I wish to thank Emery Castle and Kenneth Frederick of Resources for the Future for their continuing support of this effort.

Despite the substantial assistance received, the author takes full responsibility for the manuscript and for any errors that it may contain.

1

Introduction

One of the most significant developments in contemporary forestry is the expanding role that industrial forest plantations—forests planted, managed, and harvested for industrial wood values—are assuming in meeting the world's growing requirements for wood.

In recent decades industrial forest plantations in Europe and North America have become well established as a means of producing industrial wood, that is, wood used for the whole range of manufactured wood and wood fiber products. These regions of the Northern Hemisphere have traditionally produced the majority of the world's industrial wood.

In addition, there have been preliminary successes on many of the industrial plantations recently established in the tropics and temperate Southern Hemisphere. In these regions, the rapid biological growth rates often achieved, the vast land areas potentially available, and the successes to date are so significant that they suggest the long-term possibility of a substantial increase in the world's supply of wood and a major restructuring of the world's forest products production and trade patterns. Development of this vast potential is important in terms of meeting the world's needs for industrial wood and for fuelwood as well. Furthermore, a strong case can be made for a substantial increase in tropical forest plantations from an ecological standpoint, as discussed by Gary S. Hartshorn in chapter 8 of this report, in order to minimize the pressures on national forests and ameliorate ecological problems and environmental degradation.

The tempo of industrial forest plantation development increased dramatically during the post-World War II period and even more so after 1960 in regions of North America, the tropics, China, and the temperate Southern Hemisphere; recent estimates place worldwide reforestation as high as 9 million hectares annually ("World Wood," 1981).

The reasons for this expansion include the following:
1. Economic growth and the increasingly important role of paper and paper products have stimulated local demand for solid wood and wood fiber, necessitating a reevaluation of forest ade-

quacy in many regions where domestic natural wood supplies had once been thought to be more than adequate.

2. Increased use of paper products, some of which require the long fiber associated with softwood trees that are not commonly found in abundance in much of the world, has generated interest in the feasibility of domestic softwood plantations in regions seeking greater economic self-sufficiency and lacking indigenous long-fiber forests.

3. The phenomenal biological growth rates in tropical and semitropical regions indicate a biological advantage in timber production that can be transformed into economic advantage as well. In fact, while experience with exotic (nonindigenous) species plantations is still limited, results thus far are so dramatic that some knowledgeable observers maintain that tropical and semitropical regions relying primarily on exotic species may eventually become dominant world wood suppliers.

4. Improved transportation systems and worldwide integration of forest product markets suggest increased regional specialization in a variety of products, including timber production, based upon underlying considerations of economic comparative advantage. Whereas wood and fiber markets were once largely limited to the producing country itself, or at most to its geographic region, today forest product markets are well-integrated world markets in the same sense that world markets exist for many agricultural commodities (Sedjo and Radcliffe, 1981).

5. The real prices of certain forest resources have been rising for decades, if not centuries, as the global stock of old-growth forest is gradually depleted and as world demand increases. The rise in real prices, together with expectations of still higher real prices in the future, provides the financial and economic incentives that encourage investments in industrial forest plantations.

As indicated above, future volumes of production from industrial forest plantations could be significant, and as a result of the restructuring of production, the impact upon certain forest product trade flows could be great. Brazil alone has the potential to alter world production and trade patterns in chemical pulp by the mid-1990s (Sedjo, 1980). Some (for example, Laarman, 1981) have suggested that New Zealand, Chile, and Brazil may all have the potential to become major wood product exporters before the turn of the century. In addition, to the extent that forest plantations become common in countries that were former importers, import substitution is possible and would relieve the pressures upon other supply sources.

For industrial plantations to have a sizable and continuing impact upon global long-term supply and thus upon world markets, at least two conditions must be met. First, the underlying economics must be favorable. While some countries might choose to subsidize elements of forest production, it is unlikely that very many countries would be willing to bear the substantial costs of large subsidies to industrial plantation production.

Second, the amount of land potentially available for plantation production must be substantial. Thus, even if a region has some lands with very great forest plantation potential, unless the total land areas available are substantial, the impact on total supply can only be small. However, a relatively small area, by global standards, of highly productive forest could dramatically alter the balance and structure of the world's industrial wood supplies. To illustrate: the total land area in forestry worldwide is about 2.8 billion hectares, while the total worldwide production of industrial roundwood in 1978 was only about 1.4 billion cubic meters (FAO, 1979). Thus the average productivity per hectare in 1978 was only about 0.5 cubic meters. High-yielding forest plantations are capable of producing 15 or 20 cubic meters per hectare per year. A modest 70 million hectares or 2.5 percent of the world's forest land capable of yielding 20 cubic meters per year could have met world demand in 1978. At the reduced output of 10 cubic meters per hectare, only 140 million hectares, or 5 percent of the world's total forested area, would have been required to meet world demand in 1978.

Objectives and Dimension of the Study

In the past few years several studies have addressed the broad question of long-run worldwide demand and supply of forest projects. The U.S. Forest Service, as the result of a legislative mandate, made fifty-year projections of domestic supply and demand for forest products (USDA, Forest Service, 1979). A necessary component of that effort was a set of fifty-year projections of U.S. forest product exports and imports (Darr, 1981). Another important study dealing with future worldwide demand and supply of major products was undertaken by the Industry Working Group of the Food and Agriculture Organization (FAO) of the United Nations (FAO, n.d.). That study took a global view in its development of projections through the year 2000 of production and trade trends by major world regions. A related effort (ECE, 1976) was undertaken by the Economic Commission for Europe under the auspices of the FAO. Although the focus was upon European forest product production and trade, the study also included an analysis of production potentials from all of the world's major producing regions. Finally, an earlier RFF study (Sedjo and Radcliffe, 1981) examined the U.S. role in the world forest economy during the post-World War II period. That study, however, largely limited itself to examining past trends in U.S. forest products trade and analyzing the recent role of the United States and North America within the world forest economy.

Within the context of the studies mentioned, a number of questions recur, including the following: What is the long-run world supply and demand situation for forest products? What will be the future pattern of U.S. forest resource trade? How will the various producing regions contribute to future supplies? What role are forest plantations likely to play in meeting future world demand for forest products? Do certain regions have an inherent competitive advantage in plantation forestry? What will be the future role of the United States and of the nontraditional producers in the future world forest economy?

This report attempts to provide answers to some of these questions directly, while providing information that can assist in forming judgments on other questions. Specifically, the purpose of this study is to investigate the potential of plantation forests in a number of regions throughout the world to produce pulpwood and sawtimber while generating acceptable economic returns. The analysis explicitly examines the ability of the plantation to produce for world markets. While it should be noted that the world trade patterns can be affected by production for domestic markets that displace imports as well as by production for the export market, it can be expected that the major wood deficit regions—Japan, the northeastern United States, and Europe—will continue as dominant deficit regions for the foreseeable future. Also, by using the world market, it is possible to avoid problems generated by various distortions that occur in domestic prices as the result of governmental policies such as price controls or subsidies. If the economics are favorable for export to world markets, they will certainly be favorable for production for domestic markets.

A working hypothesis of the study is that, as discussed below, the world is experiencing a gradual transition from reliance on old-growth to reliance on second-growth forest resources, and particularly upon industrial plantations that have been designed for the singular purpose of producing the wood resource and generating an acceptable return on the investment. The main objective of the study is modest in that there is no attempt to provide comprehensive estimates of the quantities of wood that might be forthcoming through time. Instead, the analysis is directed at one aspect of the total supply phenomenon, namely, at the economic viability of plantations in selected regions, given the price and cost conditions that existed in 1979. Chapter 2 introduces in detail the construct of the representative plantation, designed to permit this regional analysis.

As discussed in chapter 3, the methodology employed in this study involves a plantation simulation model that incorporates information on both biological growth and management costs. The model explicitly considers the costs associated with timber growing—that is, the costs of site preparation and planting and of managing and harvesting timber. It estimates shadow stumpage prices by utilizing product market prices and costs of harvesting, processing, and transporting. The basic model involves twelve supplying regions worldwide and three consuming regions. Two final products (bleached kraft wood pulp and lumber) and two primary products (pulpwood stumpage and sawtimber) are considered. The analysis recognizes variations in the wood pulp by wood fiber length, that is, long-fiber and short-fiber pulp, and differences in lumber quality by introducing differences in final prices that reflect these quality variations.

Other considerations affecting the potential economic viability of plantations, such as land acquisition costs, development costs, and biological and political risk, are introduced outside the formal model, as described in chapter 4; one of the uses of the formal model is for estimating the level of development, land acquisition, and other costs that might be incurred without destroying the economic viability of an investment.

The benchmark for the analysis is the base case, presented in chapter 5. Reflecting 1979 costs, prices, yields, and so forth, the base case utilizes the best available yield and management cost data for the representative plantations, international transportation costs based on estimates of long-term costs for large volumes, and processing costs as found in optimal-size plantations in developed countries. In addition, the base case assumes that real costs and stumpage prices will remain unchanged in perpetuity. Throughout the analysis, real discount rates of 5 and 10 percent are used.

In chapter 6, sensitivity analysis is undertaken to determine the extent to which the findings of the study would be altered (a) if harvest and internal transport costs increased or decreased by 20 percent; (b) if international transport costs were 20 percent higher than in the base case; (c) if pulp processing costs were 20 percent higher than those assumed in the base case; and (d) if there were a 2 percent per annum increase in traditional producers' sawtimber stumpage prices and an increase in pulpwood stumpage prices of 1 percent per annum in perpetuity.

Chapter 7 discusses the implications of the base case and sensitivity analysis results, including the implications for U.S. forestry and for the structure of world production and trade. And finally, chapter 8 analyzes the ecological implications of tropical plantation forestry.

It is expected that this study will be useful to a variety of users, including potential forest plantation investors, both public and private. Although the study is not a substitute for a careful, site-specific project evaluation, it does provide a regionally specific assessment of the returns to a project as well as identifying the cost components to which the returns will be highly sensitive and those that will have only modest impacts upon the projected economic returns. It also provides the investor with an estimate of the amount of land acquisition and plantation development costs that can be incurred without compromising the investment's financial feasibility.

This study provides a necessary step in the ongoing investigation of the role of the United States in the world's forest economy and into questions of long-run timber supply both nationally and globally. Subsequent work could utilize the results of this study to estimate the probable impacts upon future world forest product markets and upon the role of the various major producers of nontraditional supplies of forest products.

Regions Selected for Study

The number of actual and potential forest plantation sites is large and varied, encircling the globe. However, for purposes of this study, which deals with the economic viability of industrial plantation forests in regions that are important to international markets, suitable candidate locations were much more limited.

The choice of regions to be examined was based on the following considerations. Worldwide representation was sought, including several continents and climatic zones. Both the traditional forest-producing northern temperate regions and the tropical and temperate Southern Hemisphere areas, where neophyte production currently exists, needed to be included. Furthermore, it was required that in the particular regions selected there should be some industrial forest plantation activity indicating that an independent assessment, often by the market, had concluded that the biological/economic outlook appeared promising and that at least rudimentary biological and economic data were available.

Based on these criteria, twelve regions were chosen for study (see map, frontispiece). The first three, representing the traditional forest-producing areas of the world, are the U.S. South, the Pacific Northwest (including British Columbia), and the Nordic region (that is, Norway, Sweden, and Finland). Representing the nontraditional forest-producing areas are Amazonia, central Brazil, southern Brazil, Chile, West Africa (Gambia and Senegal), South Africa, Borneo, Australia, and New Zealand.

The following section sketches the development and extent of plantation forestry worldwide, highlighting the regions selected for study.

Plantation Forestry Worldwide and Regionally

A Food and Agriculture Organization study (1978) estimates the total area of all plantation forests—including those utilized for fuelwood and environmental protection as well as for industrial production—at 90 million hectares in the mid-1970s (table 1-1). Although this is a small fraction of the world's 2.8 billion forested hectares, it should be noted that the potential impact on world markets is the result not only of more land being converted into forest plantations, but also of increased volumes of output per land unit. These volumes are likely to be large for plantations because their locations are usually determined, at least partially, by considerations of high biological growth, and also because management practices and technology usually associated with plantations increase usable growth.

TABLE 1-1. Area of Plantation Forests by Region, circa 1975

Economic class and region	Million hectares
Developed	
North America	11
Western Europe	13
Oceania	1
Other	10
Total	35
Developing	
Africa	2
Latin America	3
Asia	2
Total	8
Centrally planned	
Europe and the Soviet Union	17
Asia	30
Total	47
Total world	90

Source: Food and Agriculture Organization of the United Nations. *Development and Investment in the Forestry Sector.* FO:COFO–78/2 (Rome, 1978).

The vast amounts of plantation area listed in table 1-1 under "centrally planned" regions of Asia and the USSR are not examined in this study, since information and data are quite scarce, and since centrally planned economies typically do not use or respond to price signals for either production or marketing decisions.

Plantations in Traditional Forest-Producing Regions

There is a centuries-old history of artificially regenerated plantation forests in traditional forest-producing temperate regions of the Old World; they developed in Europe and parts of Asia in response to reduced availability of natural stocks and to rising real prices of timber. The process of replacing natural forests with managed plantations is most advanced in Europe, where plantation forests have been maintained for hundreds of years, and continue to be established today. In addition, in some parts of Europe areas are being afforested that have been devoid of forests for centuries. For example, the United Kingdom is currently afforesting in plantations some 30, 000 hectares per year.

Among European nations, the Nordic countries of Norway, Sweden, and Finland are major surplus areas for forest products, with large aggregate production and with a very large fraction of the production finding its way into major world markets, primarily in continental Europe. This has been the case at least since the inter-World War period. With the exception of the Nordic countries indicated, Europe is a very large net importer of forest products; the international potential of any European region to market forest products outside of Europe appears small.

In North America, the other major industrial wood-producing temperate region of the world besides Europe, managed plantations are a relatively recent phenomenon. In earlier periods of U.S. history when the population was moving westward, the depleted agricultural lands left behind in the East and Southeast typically reverted to forests, first in pine, and subsequently in mixed hardwood stands. (These second forests represent natural reforestation and cannot be viewed as plantations.) Later, as further reductions in the old-growth inventory occurred, plantation investments began. Although some plantation activity occurred as early as the 1930s, it was only after World War II that industrial forest plantations, almost all using indigenous species, began to become significant in the United States. During this period, artificial reforestation, together with afforestation on submarginal agricultural lands, became increasingly important. Today the vast majority of the South's forest production is consumed within the United States, with major interregional flows going to the northeastern and north central regions of the country.

Only since the 1960s have industrial plantations spread from the U.S. South to the Pacific Northwest as old-growth inventories declined. In Canada plantation forestry is still in its fledgling stages, in part because of the very large inventories of natural forests still available. The Pacific Northwest (including British Columbia) today is the single most dominant wood-exporting region in the world, with products flowing to all the major world markets (to Japan, the northeastern United States, and Europe).

Table 1-1 estimates that the total area in plantation forests for Western Europe and North America in the mid-1970s was 13 million hectares and 11 million hectares, respectively.

Plantations in Nontraditional Forest-Producing Regions

The transition to forest plantations, which mostly involve indigenous species, in the Northern Hemisphere temperate regions is notable. However, the rapid increase in the tropics and temperate southern reaches of the southern Hemisphere in land area in forest plantations, particularly those utilizing the high-yielding exotic species, could, as indicated, have profound implications internationally.

In the nontraditional forest-producing regions, the introduction of industrial forest plantations proceeded slowly at first. Brazil utilized wood from fast-growing eucalyptus plantations as fuel for its railroad systems around the turn of the century, and New Zealand and Chile both developed plantations of exotic pine well before World War II. And in the pre–World War II period, industrial plantations, also utilizing exotic species, were established in Africa and parts of Asia. But on a world scale these activities were quite modest indeed. After World War II, and particularly in the 1960s, as has been stated, the pace of forest plantation development increased greatly in these regions as well as in North America.

Of all the nontraditional forest-producing areas, South America is generally agreed to have the greatest potential for plantation forestry because of favorable biological and locational considerations and availability of vast land areas. And within South America, no region has been so active in recent years, nor has shown more potential, than Brazil (Sedjo, 1980). The level of activity in Brazil represents in part the government's commitment to reforestation. The Brazilian potential reflects the combination of vast available land areas, high biological productivity, and a favorable location vis a vis major world markets.

Chile also has the potential to produce wood volume well in excess of domestic requirements, and so is also a potential supplier to the world market. Plantations of pine have been in place in Chile for some fifty years. More recently, planting activity in the country has increased substantially, with a governmental goal of 80,000 hectares per year, and within the past five years Chile has begun actively to enter world markets.

Significant South American plantation activity is also under way in Colombia, Argentina, Venezuela, and other countries. In Venezuela, serious interest in plantations has developed within the past decade or less, and the level of activity, involving hundreds of thousands of hectares, appears to have great potential because of the availability of large land areas with few alternative uses and a location astride a major navigable artery, the Orinoco River, providing access to the sea.

The extent of industrial forest plantations in tropical and subtropical Central and South America as of 1975 is estimated by Lanly and Clement (1979) to be 2.8 million hectares, with a notable projected increase by 1980, only five years later, to 4.1 million hectares, and a projected 10.7 million hectares by the year 2000 (table 1-2).

Industrial forest plantations in the Southern Hemisphere also include those in New Zealand and Australia. The former currently has about 1 million hectares in plantation forests, consisting mostly of North American conifers. New Zealand is adding to these at a rate of some 50,000 hectares per year, and has become a factor in Pacific Basin markets. Australia's southeastern coastal area has also introduced exotic-species plantations, particularly in conifers, on a large scale, to provide for domestic long-fiber requirements.

In Southeast Asia, plantation forestry, some of which is industrial, has been undertaken in Indonesia, the Philippines, and Malaysia.

TABLE 1-2 Area of Industrial Plantations—Tropical and Subtropical America, Africa, and Asia
(thousand hectares)

Region	1975	Projected 1980	Projected 2000
Central and South America	2,786	4,128	10,705
African south of Sahara[a]	997	1,248	2,180
Developing Asia[b] and the Far East	2,892	3,719	8,265
Total	6,675	9,095	21,150

Source: J. P. Lanly and J. Clement. "Present and Future Natural Forest and Plantation Areas in the Tropics." *Unasylva* vol. 31, no. 123 (1979) pp. 12–20.
[a]Excluding South Africa.
[b]From Pakistan east excluding the People's Republic of China, Mongolia, and Japan.

Among the commercial concerns are specialized, teak plantations. There are also industrial plantations, which are of particular interest, in Kalimantan, the Indonesian part of the island of Borneo, whose establishment has followed commercial logging operations in tropical forests.

There are now plantations of various types and sizes throughout the African continent, with industrial plantation activity in East Africa (Kenya and Tanzania); in South Africa there are large plantations of various species, especially pine; and in various countries of West Africa, including Nigeria, Gambia, and Senegal, there are plantations of eucalyptus, gmelina, and tropical pine. Table 1-2 provides estimates of total area in industrial plantations in Africa, excluding South Africa.

Finally, it should be noted that in addition to the countries mentioned so far, many others have some form of plantation forests, although often not of the commercial type. There are such plantations, for example, in Japan (some 9 million hectares), Korea (4 million), India (4 million), and parts of the Middle East. And other countries such as Turkey and Iran have undertaken plantation activities with a view to commercial industrial outputs.

References

Darr, David R. 1981. "U.S. Exports and Imports of Some Major Forest Products: The Next Fifty Years," in Roger A. Sedjo, ed., *Issues in U.S. International Forest Products Trade* (Washington, Resources for the Future).

Economic Commission for Europe (ECE). 1976. *European Timber Trends and Prospects: 1950–2000* (Geneva, FAO/ECE).

Food and Agriculture Organization of the United Nations. Undated. "FAO World Outlook for Paper and Paper Board: Phases I-V," (Rome, FAO).

———. 1978. *Development and Investment in the Forestry Sector*. FO:CEFO-78/2 (Rome).

———. 1979. *Yearbook of Forest Products: 1968–79*, (Rome, FAO).

Laarman, Jan. 1981. "Discussion," in Roger A. Sedjo, ed., *U.S. International Forest Products Trade* (Washington, Resources for the Future).

Lanly, J. P., and J. Clement. 1979. "Present and Future Natural Forest and Plantation Areas in the Tropics," *Unasylva* vol. 31, no. 123, pp. 12–20.

Sedjo, Roger A. 1980. "Forest Plantations in Brazil and Their Possible Effects on World Pulp Markets," *Journal of Forestry* vol. 78, no.1 (November), pp. 702–705.

Sedjo, Roger A., and Samuel J. Radcliffe. 1981. *Postwar Trends in U.S. Forest Products Trade* (Washington, Resources for the Future).

USDA, Forest Service. 1979. *An Analysis of the Timber Situation in the United States: 1952–2030*, Review Draft (Washington).

World Wood. 1981. *1981 World Wood Review*, vol. 22, no. 8.

2

The Representative Plantation

Conceptually, the forest plantation is an economic agent charged with the task of utilizing inputs and technology in an economically efficient manner to produce a set of outputs with the goal of maximizing the plantation's economic profit. For this study, the industrial forest plantation encompasses the plantation and subsequent management of the land for wood output. The focus is on the stream of management inputs, their cost and timing, and on the physical outputs, their value and timing. While an industrial forest plantation need not be part of an integrated wood-processing operation where the woodlands and mills are operated in concert, for purposes of this study it is assumed that the plantation is part of a broader, integrated forest company.

Industrial Plantation Decision Variables

Before introducing the construct of the regionally representative forest plantation, which is designed to permit systematic examination of the twelve selected regions to produce pulpwood and sawtimber while generating acceptable economic returns, certain aspects of plantations in general need to be considered. Specifically, these involve interrelated decision variables that are tied to profit maximization on any industrial forest plantation—they include choice of location, choice of mix of outputs and level of output, and selection of management practices from among the array of technical options available.

Choice of Location

The ability to control forest location permits selection of lands where forests are the highest and best use, while lands with higher-valued alternative uses can be utilized for other purposes, such as agriculture, wilderness, and recreation. With plantations, commercial forests would tend to be established in regions of high biological productivity and good access to markets, provided this was the highest-value use.

Choice of plantation location also increases flexibility in developing a forest land configuration appropriate to efficiently providing a wood feedstock to the processing mill. If they are properly located, high-yield plantations would require less land area to continuously service a given processing complex, thereby reducing road-building and local transport costs. In addition, sites can be selected on the basis of desirable characteristics in terms of terrain and other cost-reducing features. Also, control of forest location allows the introduction of forest plantations into some regions where natural regeneration is poor or on lands that have not been forested previously.

Choice of Mix and Level of Output

As has been stated, the objective of the industrial forest plantation is to produce a set of outputs in a manner that will maximize net revenue or economic profit. In the context of the long time horizon, this is identical to maximizing the present net value. To achieve this, two related but somewhat different sets of decisions must be made by the plantation firm. The first involves the mix and level of output to be produced. The second set of decisions relates to how this desired level of output should be produced, that is, to the mix of inputs and technology that should be applied.

The forest plantation differs in a number of important respects from the firm typically discussed in the intermediate economics textbooks. The forest plantation is dealing with a time frame that is quite different from most firms. In addition, the forest plantation is a multiproduct firm capable of producing a variety of products jointly, such as pulpwood, sawlogs, peeler logs, and poles. Certain interrelationships exist between these outputs. For example, almost all sawlogs, peeler logs, and poles could be used as pulpwood. However, an asymmetry exists, since all pulpwood could not be used for sawlogs. Thus, while an economically rational firm may choose to utilize all of its stumpage for pulpwood given some set of prices, a firm choosing to produce lumber generally will not discard its residuals and thinnings. Instead, the sawlog residuals of lumber production either become inputs into a pulp mill that is part of the larger integrated firm, or the residuals are transferred through markets to separately managed pulp mills. The forest plantation therefore must decide whether to specialize in pulpwood or to produce solid wood products using the residuals as a pulpwood by-product.

The choice of mix of out-puts is, of course, constrained by the technical options available to the forest plantation and, within those constraints, is determined by the economics associated with the production of the various outputs, the markets, and the market prices of the various outputs. Further, the options available typically decrease as the rotation cycle proceeds.

Choice of Management Practices

Although not necessarily restricted to plantations, forest management and technological improvements are a major source of increased productivity, and the full range of management practices and productivity increases is possible only in a plantation setting. Forest management practices contribute to both higher biological and economic productivity. Measures to enhance biological productivity include the ability to introduce genetically superior seedlings, choose species, control spacing, limit vegetative competition, and use fertilization.

The economic productivity of the forest is also enhanced by the ability to control the quality of the output. Practices that improve the quality of output include choice of species; genetic improvement that results, for example, in straighter tree stems and increased disease resistance; and also a variety of silvicultural practices that improve stumpage quality. The latter include thinning which, while not increasing total growth, concentrates growth on fewer stems, thus producing higher-valued, larger logs. Thinning also allows higher economic productivity through the capture of mortality—trees removed by thinning typically have economic uses, whereas if left unthinned many of these trees would die as a result of natural competition.

The technical options available to the plantation are determined by the state of knowledge and the physical location of the plantation. The choice of a species, within the range of viable alternatives, further constrains the availability of technical options. A variety of technical forest management decisions must be made early in the rotation. These include the method of site preparation, choice of seedlings, density of planting, early use of fertilizers, and mode and frequency of vegetation control. Most of these decisions are made either as part of the planting decision or shortly thereafter, and to a large extent can be independent of the critical decision about whether to specialize in pulpwood or to broaden production to encompass a wider array of products.

Subsequently, however, the technical decisions that must be made have an important influence on the composition and nature of the final product. These include decisions as to thinning and length of rotation. For example, the introduction of periodic thinnings and the extension of the rotation are management practices typical in a regime directed toward a sawtimber final harvest.

Thus management decisions occurring later in the rotation cycle often reduce flexibility, although these decisions are not wholly irreversible. For an unthinned forest a late thinning may be introduced and the sawtimber values may be regained, albeit only after sustaining substantial losses as a result of biological response and time costs. At the time of final harvest, however, the decision is simply to harvest for the highest value use or to defer the final harvest. For example, even if a sawtimber forest is available to harvest, the pulpwood stumpage values may be higher. If this is the case, the harvest will either be directed to pulpwood or it may be delayed if the expectation is that sawtimber stumpage prices will rise sufficiently to make a delayed sawtimber harvest attractive.

Viewed another way, there are *ex ante* and *ex post* decisions as to the nature of the final product to which the wood will be directed. The *ex ante* decision relates to the type of forest and final harvest desired based upon expected costs and prices. (As noted, the final decision may be delayed well into the rotation.) The *ex post* decision is related to the end use of wood when the seller is faced with an actual market situation. The *ex post* decision ignores prior management expenditures and searches for the highest-value market for which the plantation output is suitable.

To summarize, the decision process is complex, with stumpage prices determining the mix of final output. However, this mix is not independent of the array of technical management options available and the costs of choosing those various options. The economic problem with respect to the set of decisions required for the plantation is, simply stated, to choose simultaneously the set of outputs and management practices that maximizes the present net value of the stream of net returns generated by the plantation.

Design of the Representative Plantation

Firms that operate existing industrial forest plantations in any region of the world have made decisions on the factors discussed above. Site location, species, mix and level of output, and management regime reflect each firm's choice from among available options for maximizing profits for their particular firm. However, examination of existing plantations, which have unique, site-specific characteristics, would not serve the purpose of broader regional analysis that this study undertakes.

Therefore, to permit systematic examination of the potential of the twelve selected regions to produce pulpwood and sawtimber while generating acceptable economic returns, a representative plantation has been developed for each region. The plantations used in this study are constructs. They are hypothetical. They do not represent any particular existing plantation run by a specific firm. Indeed, they were formulated to level intraregional differences in biological conditions and management regimes, and to avoid unique, site-specific features of any particular plantation. They are designed to be regionally representative of a "high" but not "exceptional" plantation site.

Thus these representative plantations are in a sense abstractions—that is, they are not locatable on a map. They were designed, however, to be representative of their respective regions. As such, the design of each incorporates the same essential regional biological characteristics, such as soil quality, water availability and drainage, latitude, elevation, and so forth, that would determine the inherent biological potential of any actual plantation within a given region.

To develop the representative plantations, each region's biological conditions were combined with a management regime—that is, a particular set of management practices—appropriate for that region, as discussed in the next section. Inherent biological potential together with relevant technological options can be thought of as the production function of the representative plantation. Since each plantation is based on biological features of its own region, each has a different inherent production function. The regional production functions are reflected in the biological growth rates, yields, species, and rotations used in this study's specification of the various representative plantations. The management practices specified for each representative plantation ultimately determine where on the production function the plantation will operate, that is, the mix and level of physical inputs and the level of outputs. The costs assigned to the inputs allow calculation of the management costs associated with each management regime.

Mix of Outputs and Management Regimes

As discussed earlier in this chapter, management of a forest plantation involves interrelated decisions regarding (a) the mix of outputs that will be produced and (b) the management regime that will be applied to generate that mix. For purposes of this study, two mutually exclusive mixes of output from each representative plantation are examined. The first is exclusively a pulpwood regime, while the second involves both pulpwood and sawtimber and is called an integrated regime. For the integrated regimes, thinnings are generally treated as pulpwood, while final harvest clear-cut is treated as sawtimber. In addition, residuals that would result from sawmill operations are treated as pulpwood. For each representative plantation, the mix of output is fixed. However, the mix varies across regional plantations depending upon the peculiarities of the region, the species under discussion, and the age, and hence log size, of the harvest.

The economically efficient management regime will vary depending upon the expected prices of the output, the prices of the various inputs, the discount rate, and so forth (Hardie, 1977). However, to allow tractability, this study assumes that a unique management regime associated with a particular output mix is given. The regional management regimes used in the study reflect rotation ages and other practices of large, private industrial forest plantations. Since it is assumed that these firms use profit-maximizing practices, their management regimes should approximate the economic optimal. In regions where plantations are still in their fledgling stage, a judgment was made as to the type of regime that appears to be economically optimal based upon feasible alternative practices.

In summary, the fully developed representative plantation includes the combination of regional biological potential together with specified management inputs introduced into the production process, the costs associated with each management practice, stumpage prices, and estimated yields. Tables 2-1 and 2-2, included here for illustrative purposes, exemplify a fully developed representative plantation. Table 2-1 is the pulp-

TABLE 2-1. **North America, U.S. South, *Pinus taeda*, High-Yield-Site Pulpwood Regime: Management Practices, with Associated Costs, Stumpage Prices and Yields**

Year	Practice	Cost (per hectare)	Stumpage price (per m^3)	Yield (per hectare)
0	Site preparation	$180.75	—	—
0	Planting	83.42	—	—
0–30	Stand protection	2.00 ea. yr.	—	—
4	Hardwood control	69.52	—	—
10	Controlled burn	11.63	—	—
15	Pulpwood commercial thin	—	14.94	35.35 m^3
20	Pulpwood commercial thin	—	14.94	54.79 m^3
25	Pulpwood commercial thin	—	14.94	71.23 m^3
30	Pulpwood harvest	—	18.94	373.65 m^3

Sources and Notes:
Site preparation, planting, and hardwood control costs from Dutrow (1978). Hardwood control costs are cited as "cleaning and release." 1978 "medium" costs for Southeast updated to 1979 using 1978–79 WPIs, International Monetary Fund (1980, line 63). Stand protection costs estimated from Bellinger (1981) and Weyerhaeuser (1981). Prescribed burn costs average of Weyerhaeuser (1979) and Mills and Cain (1978).
Regime from Coile and Schumacher (1964).
Yields are from Coile and Schumacher (1964)—Coastal Plain, oldfield, site indices 60 and 70 at age 25 for medium and high sites, respectively, 800 trees per acre at age 5. A conversion factor of 78.3 cubic feet per cord was used, which produced yields that are an approximate average of those published in Burkhart and coauthors (1972), Coile and Schumacher (1964), Feduccia and coauthors (1979), Smalley and Bailey (1974), and Smith (1976) (with unpublished yield tables).

TABLE 2-2. **North America, U.S. South, *Pinus taeda*, Average-Yield Site, Integrated Regime: Management Practices, with Associated Costs, Stumpage Prices, and Yields**

Year	Practice	Cost (per hectare)	Stumpage price (per m^3)	Yield (per hectare)
0	Site preparation	$180.75	—	—
0	Planting	83.42	—	—
0–35	Stand protection	2.00 ea. yr.	—	—
4	Hardwood control	69.52	—	—
10	Controlled burn	11.63	—	—
17	Pulpwood commercial thin	—	14.94	49.31 m^3
22	Pulpwood commercial thin	—	14.94	49.31 m^3
30	Pulpwood commercial thin	—	14.94	33.43 m^3
30	Sawtimber commercial thin	—	27.88	15.89 m^3
35	Pulpwood harvest	—	18.94	23.73 m^3
35	Sawtimber harvest	—	31.88	262.26 m^3

Sources and Notes:
Site preparation, planting and hardwood control costs from Dutrow (1978). Hardwood control costs are cited as "cleaning and release." 1978 costs for Southeast updated to 1979 using 1978–79 WPIs, International Monetary Fund (1980, line 63). Stand protection costs estimated from Bellinger (1981) and Weyerhaeuser (1981). Prescribed burn costs average of Weyerhaeuser (1979) and Mills and Cain (1978).
Regime from Coile and Schumacher (1964).
Yields are from Coile and Schumacher (1964)—Coastal Plain, oldfield, site indices 60 and 70 at age 25 for medium and high sites, respectively, 800 trees per acre at age 5. A conversion factor of 78.3 cubic feet per cord was used, which produced yields that are an approximate average of those published in Burkhart and coauthors (1972), Coile and Schumacher (1964), Feduccia and coauthors (1979), Smalley and Bailey (1974), and Smith (1976) (with unpublished yield tables). Yields allocated to sawtimber and pulpwood yield roughly in the same proportion as that found in Mills and Cain (1978).

wood regime for the representative plantation of the U.S. South (high-yield site), and table 2-2 is the integrated regime (average-yield site) of the same region. For tabular presentation of all representative plantations developed for this study, see appendix A.

Data Sources and Quality

The data used to develop the representative plantations were derived from a wide variety of published and unpublished sources, the latter including representatives of plantation firms, a national paper and wood products association, and a university silvicultural department. These sources are specified in the notes for the table of each representative plantation in appendix A. As indicated below, the quality of the data varies considerably, but in each case the data are the best available to the scholarly community.

Much of the data is excellent and well documented, especially in the cases of the U.S. South and Pacific Northwest. For these two regions, in fact, so much information was available that two alternative site classes—high-yield and average-yield sites—were analyzed. (Thus, appendix A provides four tables both for the U.S. South and for the Pacific Northwest: they are for an average-yield site with a pulpwood regime, an average-yield site with an integrated regime, a high-yield site for a pulpwood regime, and a high-yield site for an integrated regime.) The determination of which site classes would best represent the high-yield and average-yield sites was judgmental.

The data for the Nordic forests, as well as for Australia, New Zealand, and South Africa, appear to be of high quality and reasonably representative of the costs, yields, and regimes that occur in those regions.

By contrast, in the case of Brazil in particular, well-documented yield curves for exotic species were not located for any of the three Brazilian regions investigated. However, personal interviews with many knowledgeable people, such as representatives of the Associacão Naciónal dos Fabricantes de papel e Celulose, provided a consistent view of typical yields for exotics for the three regions, which were incorporated into the study. In assessing the costs and regimes, the data used were those determined by the Brazilian government as being typical of plantation forestry. While these cost estimates have been used to determine reimbursement to the private investor in Brazil as part of the incentive program, they also provide documented estimates of the levels and time profile of costs likely to be incurred in plantation forestry. Some have argued that the incentives program is overly generous in its assessment of costs, but to the extent that these costs are high, the present net values (PNVs) estimated in the study for Brazil err on the low side. Our judgment is that the plantation data for Brazil are reasonably accurate and that they are not so far in error as to markedly affect the PNV estimates.

The data for certain other regions were also difficult to obtain, and sometimes their accuracy was suspect. For example, while the data for West Africa were obtained from government sources and were published in a consulting report, the differences between the plantation costs of gmelina and eucalyptus, the two species examined, have not been adequately explained. Nevertheless, West Africa was retained in the study, since the costs do not deviate greatly from the range of cost estimates obtained for other regions. By contrast, the data from Chile, while not plentiful, are viewed as reasonably accurate. Yields, regimes, and costs are well documented.

Perhaps the weakest data are for Borneo. The data on yields and regimes are based on conversations with members of the Weyerhaeuser Company at their forest concession in Indonesia. Since there was only four years' growth experience, the yields were necessarily "best guesses." The regimes represent judgments as to the sets of activities that are likely to be appropriate and economic. As noted in the Borneo tables in appendix A, the costs were drawn from experience in the Amazon. The rationale for using Amazon cost data was that the species used were the same, both regions were tropical, and both regions were located a considerable distance from the country's centers of population and industry. Nevertheless, it is obvious that the results from regions like Borneo, with very weak data bases, must be interpreted with a great deal of care.

Reference

Hardie, Ian W. 1977. *Optimal Management Plans for Loblolly Pine Plantations in the Mid-Atlantic Region* MP906 (College Park, Agricultural Experiment Station, University of Maryland).

3
Forest Plantation Simulation Model

The forest plantation simulation model developed for this study was designed to be as simple as possible while still allowing for the investigation and comparison regionally of the economic returns to industrial forest plantations. For purposes of this study, it was necessary to capture the essential features that affect the relative competitive position of each regionally representative plantation. Therefore, the model focuses on the links between the world market and the representative plantations, stumpage prices, and ultimately on costs and returns. In essence, the model is a way of systematically tying the regionally representative plantations to the world market and of examining the stumpage price implications. It should be noted, however, that the model is not a complete general equilibrium model and is not intended to be a full representation of the world industrial wood market. Rather, it treats the world price of final products as being generated externally (exogenously) by forces not specified in the formal model, in much the same way that the firm in a competitive market views the market price as exogenous. Of course, this does not imply that world prices will not change over the period examined by the model. Instead, it assumes that changes in world prices would be viewed by each representative plantation as independent of its own individual actions. Therefore, to take into account the possibility that market prices may change in a systematic, predictable manner over the period of investigation, the study includes sensitivity analyses that allow for changes in the exogenous price.

Over the long time period (as much as eighty years for one particular rotation) examined in the model, it might be expected that some of the assumed relationships would break down. For example, for purposes of logical completeness and also to reflect the 1979 situation, the final product prices in the three components of the world market are tied together via transport costs from the Pacific Northwest region. The old-growth inventories of the Pacific Northwest and British Columbia are of such magnitude that the dominance of this region may be expected to persist well into the twenty-first century. Eventually, however, this relationship is likely to erode. Particularly to the extent that

Southern Hemisphere plantations are successful, the role of the Pacific Northwest might decline somewhat sooner. Nevertheless, changes of this type in the structure of the world market will not seriously affect the results of this study, since they are limited by transport costs and arbitrage, that is, the assumption of a reasonably competitive world market precludes large changes in the structure of market prices across regions. Thus, only relatively small changes in stumpage prices and present net values (PNV) would result from the effect on market price structure of a diminished role for the Pacific Northwest. Changes in the exogenous price, by contrast, would have substantial effects upon the economics of the various representative plantations. Some of these effects are examined in the sensitivity analysis undertaken in this study, and further sensitivity analysis could be pursued if deemed desirable.

Basic Components of the Model

The forest plantation simulation model discussed in this chapter is presented mathematically in appendix B. The analysis focuses on the cost and receipt differences in the representative plantations as generated by their respective production functions (that is, biological potential as enhanced by use of technological options), management regimes, management costs, and international transport costs based on the varying distances to the principal world markets. The model determines which of the major world markets—Japan, the northeastern United States, or western Europe—would be a representative plantation's most profitable foreign market.

Utilizing the differences in the production functions and international transportation costs, the model allows for the determination of implicit or shadow stumpage prices for each representative plantation, as discussed later in this chapter. A Faustmann-type approach (Gaffney, 1957; Bentley and Teeguarden, 1965; Gregory, 1972) as incorporated into a computer program relates the shadow stumpage prices to output, management costs, and the discount rate to calculate the present net value per hectare for the various plantations. An internal rate of return (IRR) is also calculated.

Present Net Value Criterion

Present net value of a perpetual series of plantations is the principal investment criterion used for this study. It is calculated using the Forest Service computer formulation of a Faustmann-type model developed by Goforth and Mills (1975).[1] This approach subtracts the discounted costs of the various management activities from the discounted receipts to give a present net value, what in Faustmann terms would be the capitalized value of the discounted "soil rents" in perpetuity. The costs are drawn from the representative plantations, while the stumpage receipts are the product of the stumpage harvest volumes and their shadow prices as determined by the model. Two real (inflation-free) discount rates are used, 5 percent and 10 percent, with the principal focus of the study upon the results for the 5 percent discount rate, since real long-term interest rates appear to be about 5 percent.[2]

---

[1] A few minor changes were introduced into the computer model, for example, a modification to allow for differing rates of long-term price growth for pulpwood and sawtimber stumpage. The computer model uses as inputs the management costs and physical outputs presented in appendix A, together with the shadow stumpage prices developed for the various sensitivity analysis scenarios, as presented in appendix D.

The model does not solve for the determination of the optimal rotation length given the prices, costs, discount rate, and biological growth. Therefore, throughout the analysis the rotation length has been assumed to be a constant and was that length commonly used, or, alternatively, the rotation length viewed by knowledgeable observers as economically optimal. Economic rotation length will vary according to conditions, since economically optimal rotation length is defined as that which maximizes the PNV given the various costs and receipts and their respective time profiles in perpetuity.

Although these simplifications undoubtedly introduced errors, we believe these errors to be small and the estimates obtained therefore to be useful indicators of the values generated by rotation of economically optimal length.

[2] For a nontechnical discussion of the appropriate discount rate applied to forestry, see Walker (1981); and Row, Kaiser, and Sessions (1981).

In addition, internal rates of return are calculated for each plantation. Since this criterion favors short payback projects, it generates a somewhat different ranking for projects, tending to view more favorably the short-rotation pulpwood project over the longer-rotation integrated pulpwood/sawtimber operation. This study, however, will stress the results given by the present net value criterion, because it is well established in the economics literature that the present net value criterion is theoretically superior to the internal rate of return approach (Hirschleifer, 1958). In addition, the focus upon present net value allows for an analysis of the magnitudes of development and land acquisition costs that might be reasonably incurred.

The model consists of five other basic components in addition to the computation of present net value (see figure 3-1). They are: (1) a world market, which integrates the different consuming regions; (2) an international transport cost matrix linking the various producing regions' plantation processing facilities with the world consuming market; (3) a processing complex consisting of pulp mills or sawmills or both, which are located in each of the timber-growing regions; (4) a harvest and internal transport component; (5) the set of regionally representative plantations.

As shown in figure 3-1, components 1, 2, 3, and 4 are used to derive implicit or shadow stumpage prices for both pulpwood and sawtimber. The Faustmann-type model discussed above combines the relevant cost and receipt data of the regional plantations (component 5) with a discount rate to estimate the present net value and internal rate of return of the various regional plantations. The following sections discuss components 1 through 4.

World Market Component

Discussion of the "world market"—the model component needed to integrate various major markets into a single, efficient market—involves identification of both the relevant commodities and the relevant regions.

First, regarding commodities for this study, two basic homogeneous goods—wood pulp and sawnwood—are assumed to be traded. They were selected because both are involved in large-scale international trade globally, and because both commodities generally have small or no tariff duties. In addition, different final products are introduced to reflect quality differences, for example, long- and short-fiber wood pulp and high- and low-quality lumber (see section on "Choice of Final Products" in this chapter). It is assumed that within the world market, a single price prevails everywhere for a homogeneous good, except to the extent that there are differences in international transport costs (see next section). This is because, to the extent that a temporary disequilibrium might result in price differences in excess of those accounted for by transport costs, arbitrage will occur, that is, an increased flow of products will move to the high-price regional market, depressing the price and restoring the equilibrium.

Regarding the relevant regions in the world market component, an examination of international trade flows in forest products reveals Japan and Western Europe to be the principal world deficit regions. To this we might add the northeastern United States (including the north central region), which receives high volumes of wood inflows both from other regions in the United States and from eastern and western Canada. The dominant world surplus region is the Pacific Northwest/British Columbia (PNW/BC) region of North America. It is the only region of the world that simultaneously provides significant volumes of both wood pulp and lumber to all three major subregions of the world market (Sedjo and Radcliffe, 1980).

The PNW/BC region provides a strong net flow of forest products, including sawnwood and bleached kraft pulp, to East Asia—particularly Japan—and it also provides a flow to both the East Coast of North America and to Western Europe. The East Coast of

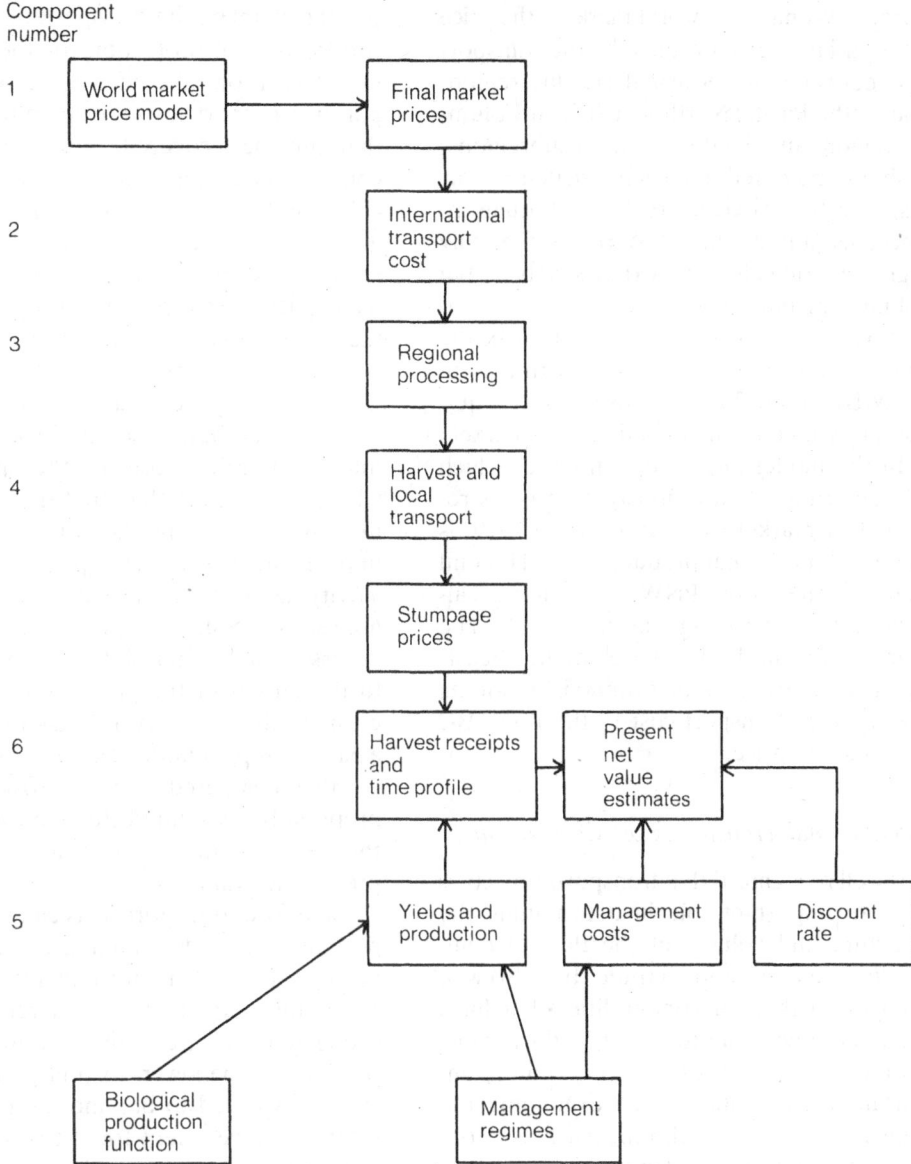

Figure 3-1. Model components.

North America, particularly the Canadian maritime provinces, provides additional net forest product flows to Europe. Hence the dominant world forest products trade flow can be viewed as emanating from the West Coast of North America and terminating in Japan, the northeastern United States, and Western Europe. Important but somewhat less massive "secondary" trade flows exist, for example, those from Southeast Asia to Japan, and to a lesser extent, to the United States and Europe, as well as large flows from the Nordic countries to the rest of Europe and from the Soviet Union to both Europe and Japan.

Consistent with the preceding identification of the major world wood deficit regions, this study defines the relevant world market consuming subregions as being Japan, the northeastern United States, and Western

Europe. Within this "world market" the price differentials are determined by the transport cost between the principal surplus region, that is, the Pacific Northwest/British Columbia region, and the three market subregions. It should be noted that such a structure corresponds to the existing real-world situation, since, as stated, the PNW/BC is the only region worldwide that exports substantial volumes of both wood pulp and lumber to all three of the world market outlets in the study.[3] Thus also, marginal adjustments in PNW/BC trade flow can ensure price equilibrium in all components of the world market.

In the model the European price, which was determined exogenously, that is, as revealed by market transactions circa 1979, is the benchmark final product price. The mill shadow price in the PNW/BC region is calculated by subtracting transport costs. The market price in the Japanese and northeastern U.S. markets is determined by adding the relevant transport cost to the PNW/BC mill shadow price.

International Transport Cost Component

It is well recognized that transportation costs play an important role in determining the structure and volume of the flow of commodities in international trade. It is also well recognized that for commodities that have relatively low value-to-weight ratios, transportation costs become a relatively more important consideration when examining the ability to compete in distant markets. Both of the commodities under study, wood pulp and sawnwood, have relatively low value-to-weight ratios, and thus transportation costs to distant world markets can be a crucial component in the final determination of the degree of competitiveness of the commodity in world markets. From the view of the importing country, the transportation costs afford a form of protection for the domestic industry. From the viewpoint of the exporter, transport costs are an additional cost that must be recovered if the producer is to compete successfully in the export market.

Within the methodological approach used in the study, the role of transport costs is critical, since it is these costs that determine the implicit or shadow stumpage prices within the various regions. Differences in transport costs between regions will be reflected in differing stumpage values, because it is assumed in the base case that processing and internal transport costs are the same everywhere. Thus the differential transport costs ultimately determine the differentials in the implicit price of the stumpage. In the sensitivity analysis scenarios these assumptions are relaxed. Stumpage prices, of course, are necessary in the analysis for assigning a value to the output of the plantation in order to estimate the gross revenues associated with a particular plantation. These gross revenues are then compared with the gross costs, appropriately discounted, to give estimates of the internal rate of return and discounted present net value.

Therefore, transport costs can play an important role in determining the returns to the investments in forest plantations. The closer the plantation and its related processing activities are to the final market, *ceteris paribus*, the lower the transportation cost can be expected to be, and the greater the return on investment for the forest plantation activity. As the overall level of transport costs declines, the returns to the forest plantation investments in each of the regions increase and the role of distance from market becomes less important. In the extreme, with zero transport costs, location relative to the major markets would cease to be an economic consideration. As the transport cost rises, the returns correspondingly decline and the importance of the distance from the world market increases. Thus, changes in these costs can result in a submarginal plantation's becoming supermarginal, and vice versa.

---

[3] It should be noted that Japanese tariffs on certain types of forest products distort the actual trade flows somewhat away from processed and toward unprocessed wood. However, the resulting bias will be to increase the implicit stumpage prices.

Finally, the structure of transport cost is important also in evaluating the comparative economic returns of the various regional plantations. To the extent that transport cost differentials are large across regions, the plantation's location relative to the market becomes an increasingly important factor in determining its economic competitiveness. As the transport cost differentials become smaller, comparative location becomes a less important consideration.

The methodology for the international transport cost component is discussed in detail in appendix C. Briefly, using empirical data, a separate formula is developed for wood pulp and for lumber; each treats international transport costs principally as a function of distance.[4] The international transport costs between each producing region and each world submarket are calculated. This allows the determination of a consistent set of free-on-board (FOB) mill prices for wood pulp and lumber at the various regional mills. The submarket generating the highest FOB price is defined as that region's principal market.

Processing Component

The model includes a processing component for each region for both wood pulp and sawnwood. This component assumes a technically efficient bleached kraft pulp mill and a sawmill (or set of sawmills) of technically efficient size, as reflected in mill cost in the United States and the Nordic regions. In the absence of systematic variations in costs, the base case assumes that the total cost of these mills is identical for all twelve regions. Processing costs enter the analysis as a dollar cost for the value added that is produced by the mill. Thus, for example, the value added by mill processing is set at $133 per thousand board feet of lumber and $275 per metric ton (Jaako Poyry, 1977) of bleached kraft pulp. These costs allow for the calculation of a residual between final product prices and costs that is defined as the shadow price of stumpage (figure 3-2).

The assumption of similar costs for identical mills regardless of the physical location is not unreasonable. Presumably, a technically efficient mill can be purchased on the world market at very similar prices regardless of its ultimate location. While labor costs can be expected to vary by region, and particularly in the case of a sawmill, possibilities do exist for the introduction of labor-using techniques. In general, labor costs in the mill will be a fairly small fraction of total cost, particularly for a pulp mill.

There are, of course, bona fide economic and financial reasons suggesting that mill costs on actual plantations would vary by region, with inaccessible locations incurring higher construction costs. For example, the total capital costs associated with a project such as the Jari project, located 400 miles up the Amazon, are likely to be higher than average. Power and other costs may also vary by region. Finally, for inaccessible regions, labor costs—particularly for technicians required in a pulp mill—may be high both because of high salaries required to attract them and because of costs for providing amenities and social services. While some of these considerations are, to some extent, site-specific, there seems little question that some regions would experience systematically higher processing costs.

Mill costs on actual plantations may also vary by region because of market distortions that result from governmental policy. Tariff policy, by encouraging the use of high-cost, locally produced goods, bureaucratic red tape, and other such price-distorting, cost-increasing policies can systematically increase processing costs within a region. However, while these policy-induced costs are important in any financial analysis, they do not reflect the fundamental underlying economic costs associated with production in a region.

Instead of attempting to ascertain variations in processing costs between regions, this study employs a sensitivity analysis (see

---

[4]This model is similar to generic models described by Beckman (1981).

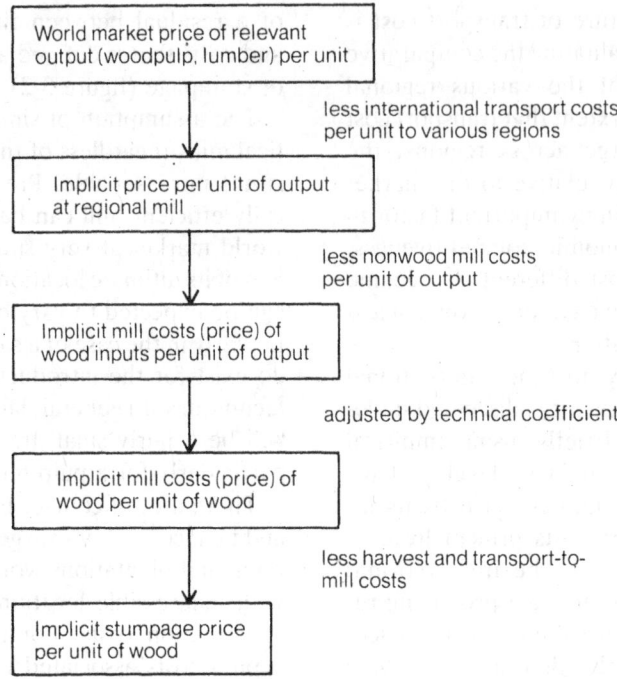

Figure 3-2. Implicit stumpage price determination.

chapter 6), which allows pulp mill and sawmill processing costs individually to increase by 20 percent in each region. The analysis provides insights as to the impact of changing processing costs regardless of whether they are the result of real economic costs or of artificial, policy-generated restrictions.

Harvest and Internal Transport Cost Component

The optimal forest plantation for minimizing harvest and internal transport costs would be situated on flat terrain, would produce large logs, would have large final harvest volumes, would be high yielding, and would have a contiguous, compact land-holding pattern of sufficient size to provide feedstock to optimal-size mills, which would be located at seaports. In addition, access to capital and well-trained, low-wage labor would, of course, be desirable.

Harvesting costs can be quite variable depending upon the factors listed above. Not only can these costs be expected to vary between regions and between plantations within the same region, but costs would also vary within a given plantation. In addition, harvesting costs can be expected to differ systematically between thinnings and final harvest, since thinning involves smaller logs, lower volumes, and harvesting procedures designed to avoid damage to the remaining trees.

Internal transport costs vary with distance of the harvest area from the mill, mode of transport, terrain, and pattern of plantations. To the extent that a plantation had well-patterned contiguous lands, the internal transport costs would be lower. In addition, a high-yielding, fast-growing plantation requires a smaller land area to support a given-size mill, so that internal transport costs would be reduced as a result of lower average haul distance.

It is assumed that the representative plantations of each region under examination consist of well-patterned contiguous holdings. The harvest and internal transport costs for this study are defined as those incurred in the process of logging and transporting those logs to a nearby mill. For the base case, these costs are assumed to be $12 per cubic meter for clear fellings and $16 per cubic meter for thinnings for all representative plantations. The sensitivity analysis relaxes this assumption and allows for alternative costs for this operation.

Stumpage Price Component

A critical element in the analysis is the stumpage price; a significant portion of the formal model is designed to calculate this implicit or shadow stumpage price, which, as has been indicated, is required in order to estimate present net value and internal rate of return.

Stumpage is defined as standing timber, that is, timber on the stump. The stumpage price, therefore, is the price of the right to harvest a particular stand of timber. For this study the stumpage price is quoted in value per unit of volume—1979 U.S. dollars per cubic meter.

Stumpage price is a residual concept. Given a processed commodity such as bleached kraft pulp, the stumpage price of the pulpwood in a competitive market is the market price of the pulp less the sum of the processing, transport, harvest, and other required production costs. In other words, subtracting the costs of processing and other inputs from the price of the final product results in a residual that is defined as the stumpage price. The procedure just described is used in this study, and is illustrated in appendix D, which presents calculation of base stumpage prices for loblolly pine in the U.S. South. This approach for the calculation of the implicit price is theoretically consistent with the economic concept of derived demand (Friedman, 1962) and is illustrated in figure 3-2. Although markets for stumpage exist, a shadow price is calculated for stumpage based upon long-term considerations. Thus the analysis abstracts from short-term cycle fluctuations that might be expected to occur periodically. Since international transport costs vary by region, interregional differences in stumpage prices are to be expected.

For forest stands, harvesting can result in positive economic benefits as long as the implicit stumpage price equals or exceeds zero. An implicit stumpage price of less than zero, however, implies that the costs of harvesting, transporting, and processing this source of wood exceed the world market price obtainable for the final product, and thus, the resource is economically submarginal.

Choice of Final Products

As already indicated, two commodities—sawnwood (lumber) and bleached kraft wood pulp—were selected as final products for this study. The question arises as to why these were chosen—why not, for example, wood chips and sawlogs, or perhaps mill paper and plywood. The reasons for the choice are several. First, experience reveals that wood pulp and sawnwood are heavily traded internationally. Second, trade duties are generally absent for trade in wood pulp and sawnwood, while they are becoming much more common at the next stage of fiber processing, such as for mill papers. Therefore, distortions from these restrictions are likely to be minimal. Finally, there is a theoretical reason for the choice of wood pulp and sawnwood. Transportation costs do afford a level of effective protection to the various stages of the production process (Finger and Yeats, 1976). Given the particularly low value-per-unit weight ratio of forest products in the production of the primary resource, the protection afforded the domestic primary resource by the high unit transport cost is likely to be large, since transport costs are likely to be a high portion of the CIF (cost, insurance, and freight) input price. However, the effective protection, that is, protection relative to the value added, afforded the next stage in processing by transportation costs is likely to be small, since the protection afforded by the transportation costs on the primary product tend to offset the effect of transportation costs on the next stage of processing.[5] Thus, for forest products, the pulping and lumber-making stages of production and trade appear least encumbered by the protection afforded by both trade restrictions and transport costs.

A case could be made for substituting plywood for the lumber used in our analysis, or alternatively for the introduction of a plywood component along with pulp and lumber. However, to facilitate the analysis, it is important that the number of commodities be limited. Thus the focus upon wood pulp and sawnwood.

WOOD PULP. Wood pulp is unaffected by consideration of taste and custom. There are, however, various qualities of bleached kraft wood pulp, which is extensively traded internationally, and is used in this analysis as surrogate for all pulps. The most important distinction relates to the fiber characteristics, and the pulps are distinguished as either long-fiber or short-fiber, depending upon the type of wood utilized for the pulp production. These differences are captured in the model by attaching a 10 percent discount to the short-fiber pulp, a discount that is reflective of the actual market price structure that has existed in recent years. There are also quality differences associated with the different species used as feedstock for the pulping process. For example, short-fiber wood pulp made from tropical pine is qualitatively somewhat different from that made from Norway spruce. However, these differences are reflected in only small market price differentials (perhaps 1 percent) and have not been introduced into the analysis.

LUMBER. Sawnwood or lumber is not nearly so well standardized as pulp. Varying amounts of processing are involved in the production of different sawnwood items. In addition, there is an important quality difference that results from the quality of the wood utilized. For example, clear lumber (knot- and defect-free) is highly valued for some purposes and is particularly highly valued within some cultures. This is true for sawnwood that is visible in a structure, but is also true in some cultures, such as in Japan, for concealed lumber, since knots and defects are perceived as structural weaknesses.

A representative basket of lumber is assumed to be produced, transported, and then transacted in the world market. A distinction is made between softwood and hardwood lumber, with the hardwood lumber (in this case gmelina or eucalyptus) selling at a discount of 10 percent. As noted, there is also a quality component associated with softwood lumber. In world markets this is reflected in species, but it also relates to such features as the incidence of knots, the number of rings per inch, the properties of the wood for construction, and so forth. These characteristics are a function of the age and species of the timber from which the lumber was made, and are often related to log size.

The stumpage price of sawtimber is also related to log size for other reasons. Generally, a higher portion of the volume of large logs than of small logs is usable as lumber, allowing for production of high value per unit of input. Furthermore, large logs require less handling per unit volume, both in

---

[5]This can be illustrated with a hypothetical example. Suppose (a) transport costs add 50 percent to the import price of logs, (b) log costs constitute 50 percent of the value added to domestic lumber production, and (c) transport costs add 25 percent to the imported cost of lumber. Now for the input of logs, the effective protection is simply the nominal protection or equivalent to an ad valorem duty of 50 percent. Using now the standard effective protection formula to examine lumber:

$$EP = \frac{T_1 - (aij) T_2}{1 - aij} = \frac{25 - (.5)(.5)}{5} = 0$$

where EP = effective protection
T_1 = transport cost as a percent of lumber imports
T_2 = transport cost as a percent of log imports
aij = percentage of lumber value added attributed to logs

In this case $EP = 0$, since the protection that transport cost afforded domestic logs (T) has entirely offset the protection created by the lumber transport costs.

harvest and in processing. Large logs frequently mean higher volumes of standing timber per unit of land, which lowers unit handling cost. Larger, older logs also tend to have the desired characteristics in terms of wood quality for lumber.

Some of the features discussed above have been captured explicitly in the model. For example, it recognizes two size ranges of logs—12-in. to 17-in. and 6-in. to 11-in. diameters—and has a different lumber/pulpwood/loss conversion factor associated with each (see appendix D). Also this problem is explored through the introduction of a sensitivity analysis scenario that provides a 10 percent market price discount for conifer lumber produced from exotic plantations in the tropics and temperate Southern Hemisphere. Treatment in this manner can be viewed either as assuming that the output of the exotic plantations is always sold at discount or as assuming that a greater portion of the output of exotic plantations is used to produce lower-quality and lower-priced lumber.

The net result of these various adjustments is that the simple two-product model has been adapted to deal with multidimensional products. The entire range of wood pulps, both long- and short-fiber, is roughly captured with grades of bleached kraft wood pulp. The wide variety of types of lumber is captured by adjustments for log sizes, species, and region of timber production.

References

Beckman, Martin. 1981. "Reflections on Palander's Beitrage nur Standortstheorie." Paper presented to the conference on Structural Economic Analysis and Planning in Time and Space, Umea, Sweden, June 1981.

Bentley, William R., and D. E. Teeguarden. 1965. "Financial Maturity: A Theoretical Review," *Forest Science* vol. 11, pp. 76–87.

Finger, J. M., and A. J. Yeats. 1976. "Effective Protection by Transportation Costs and Tariffs: A Comparison of Magnitudes," *Quarterly Journal of Economics* vol. 90, no. 1, pp. 169–176.

Friedman, Milton. 1962. *Price Theory: A Provisional Text* (Chicago, Aldine) pp. 148–155.

Gaffney, Mason M. 1957. *Concepts of Financial Maturity of Timber and Other Assets.* Agricultural Information Service 62 (Raleigh, North Carolina State College, Department of Agricultural Economics).

Goforth, Marcus H., and T. J. Mills. 1975. *A Financial Return Program for Forestry Investments.* Agricultural Handbook No. 488 (Washington, U.S. Forest Service).

Gregory, G. Robinson. 1972. *Forest Resource* (New York, Ronald Press) pp. 279–295.

Hirshleifer, Jack. 1958. "On the Theory of Optimal Investment Decision," *Journal of Political Economy* vol. 66, pp. 329–352.

Jaakko Poyry. 1977. *Jaakko Poyry Report: 1975–1977.* (Helsinki, Finland, Jaakko Poyry Consulting Oy) pp. 5–6.

Row, Clark, H., Fred Kaiser, and John Sessions. 1981. "Discount Rate for Long-Term Forest Service Investments," *Journal of Forestry* vol. 79, no. 6, pp. 367–369.

Sedjo, Roger A., and Samuel J. Radcliffe. 1980. *Postwar Trends in U.S. Forest Product Trade: A Global, National and Regional View.* Research Paper R-22 (Washington, Resources for the Future).

Walker, John. 1981. "National Forest Planning: An Economic Critique." Paper presented at RFF conference, "Coping with Pressures on U.S. Forestlands," Washington, March 30–31, pp. 34–35.

4

Extramodel Factors Affecting Investment in Industrial Forest Plantations

Chapter 3 describes the formal model developed for estimating the present net value associated with the forest management activities of an industrial forest plantation. In addition to the components of the model, which explicitly involve only the costs directly associated with timber growing, management, harvest, and transport, other factors are critically important to assessing the economic viability of industrial forest plantations. These factors include land acquisition costs, development costs (specifically, the costs associated with the introduction of the necessary physical and social infrastructure required to allow for undertaking the set of activities called a management regime), and risk. However, they do not lend themselves to incorporation into the formal model, since they vary considerably by region and site and are often extremely difficult to quantify. This chapter addresses the effects of these other factors, including their implications regarding the results presented in chapters 5 and 6.

Two questions arise with respect to land acquisition and development costs as they are treated in this study. First, does the formal model provide information useful for assessing development costs? Specifically, does the analysis provide information on the level of development costs that could be incurred by a rational investor? Second, is there any assurance that lands allocated to industrial forest plantations, even those with relatively high economic returns, are being allocated to their highest economic use? Briefly, the answer to both questions is that the formal model results, together with market information, provide the basis for rational economic assessment of industrial plantations, as is discussed in detail in the rest of this chapter.

Land Acquisition and Development Costs

In a free market with perfect information, the price of the land is simply the land's present net value less the requisite development costs for its highest-value use. In addition, there would be an adjustment downward in the market price on the land and the present net value (PNV) to account for any unusual risk associated with any particular land use. Stated another way, any economic rents could be expected to be capitalized into the market price of the land with an appropriate adjustment for risk. The present net value obtained in chapters 5 and 6 represents estimates of the price that the rational investor would be willing to pay to obtain the rights to manage and collect the economic returns generated by an industrial forest plantation. Insofar as the estimated PNVs are accurate, a large excess of PNV over land price suggests either high risk or large development costs or both. A land price exceeding PNV in forestry suggests that there is a higher-value alternative use. In general, one would expect the high-return investment site to have either high land prices, high development costs, high risks, or some combination of the above. To the extent that the market fails to perceive the true potential of a site, the market price will undervalue the land, and large returns will be experienced by the perceptive investor.

The above perspective, together with the study estimates of PNVs, provides information to the potential investor, private or public. The difference between the market price of a unit of land and the expected PNV from that unit of land is the maximum amount of development costs that can be incurred and still generate an acceptable return (adjusted for risk) to the investor.

Jari: An Illustration

Although this study does not attempt a systematic estimate of acquisition and development costs by region, it may be useful to examine illustratively a region that has generated a great deal of interest, where development costs are high. A good case in point is the Jari project in the Amazon.

DEVELOPMENT COSTS. The Jari project is a controversial private development project begun in Brazil in the late 1960s adjacent to the Jari River, a tributary of the Amazon. Although there are several commercial activities occurring simultaneously, the major feature of the project is the establishment of an industrial forest plantation. By the late 1970s, over 100,000 hectares of plantation forest had been established featuring gmelina and tropical pine.

Although we do not have any financial data other than those commonly discussed in the press, it may provide insight if we use the commonly discussed estimates to illustrate the financial costs of development. A frequently quoted figure indicates that Jari development costs are about $700 million dollars (including about $3 million for land acquisition). It should be noted that there are dimensions to the Jari investment that are related to the agricultural activities, cattle raising, and kaoline mining, as well as to forestry, and it is unclear whether the dollar figure cited refers to all expenditures, both capital and current, for all activities, or whether the costs refer only to investment costs related to the forest plantation and processing activities. For illustrative purposes let us assume the latter. It is also sometimes asserted in the press that the pulp mill and sawmill complex cost about $400 million. If this is accurate, the implication is that about $300 million was invested in development costs and plantation establishment.

The data for this study indicate that gmelina-stand establishment costs in the Amazon are about $200 per hectare, or about $20 million for 100,000 hectares. This suggests that the plantation development costs are about $280 million ($300 million less $20 million) or about $2,800 per hectare. This figure compares with an estimated PNV (at a 5 percent discount rate) for the base case of this study: as indicated in chapter 5 dealing with model results for the base case, PNV for the Amazonian representative plantation in gmelina is $3,184 per hectare for an integrated (sawtimber) regime and $2,530 for the pulpwood regime (see table 5-1). If these figures are accurate, if the plantation in the case cited above maintains itself at its current size, and if the other costs and relationships are accurately represented in our model, then the PNV of the integrated regime using a 5 percent interest rate would be sufficient to cover the plantation development costs and to generate a residual of $384 per hectare. Thus the plantation generates a modest PNV using a real discount rate of 5 percent after covering development costs. Note, however, that if the plantation must rely upon the pulpwood regime, the PNV before development costs is only $2,530, which leaves a net of minus $270 per hectare after development costs. Thus, given this information and a real discount rate of 5 percent, a rational investor would incur the illustrated development costs for a gmelina plantation only if the investor was confident that the integrated option was viable. (Note that our estimate of the PNV of the pine plantation in table 5-1 is somewhat higher than that of the gmelina).

If in the illustrative-case plantation there was eventual establishment of a total of 400,000 hectares, presumably, the subsequent development costs per hectare would be lower than initial costs, since much of the infrastructure would be in place. Suppose that the development costs associated with the additional 300,000 hectares were an additional $200 million, or about $677 per hectare. At the margin, the additional development costs would be more than offset by the PNV estimated by the model of either $2,530 or $3,184. Moreover, the average development costs of the plantation would fall to $1,200 per hectare, leaving an after-development PNV of between $1,300 and $1,800 per hectare.

Wood Pulp Processing Costs

To add another dimension to the illustrative analysis above, suppose that the wood pulp processing costs are 20 percent above those generally experienced worldwide. This could occur as a result either of high capital cost or of high variable costs. This particular situation is examined in a sensitivity analysis scenario (see chapter 6) and appears in tables 6-16 through 6-18. In this case, using a 5 percent real discount rate, PNV of a pine pulpwood plantation drops to only $262 per hectare, while the gmelina pulpwood plantation becomes strongly negative ($-$1,680). The effect on the integrated planatation is less, with the PNV becoming $1,838 per hectare for pine and $291 per hectare for gmelina. Note, however, that the development costs for the 100,000 forested hectares could not be covered for any of the species or regimes considered and that only the integrated pine regime development costs would be justified for the expanded 400,000 hectares of forest. Thus, for the illustrative case, the economic success of the plantation is dependent upon controlling the wood pulp processing costs.

Total Plantation Investment Costs: An Illustration

In this section we will examine illustratively some investment cost estimates and considerations that follow from the estimates of regional PNVs. Specifically, we will examine and estimate the development costs that could be incurred in the process of developing a vertically integrated forest plantation complex.

A technically efficient bleached kraft pulp mill has a capacity of about 260,000 metric tons per year and, with the power plant, the capital costs are about $300 million dollars. To produce 260,000 metric tons of bleached kraft wood pulp requires 1,066 cubic meters of short-fiber or 1,222 cubic meters of long-fiber pulpwood using conventional coefficients. For a forest that is yielding 20 cubic meters per hectare per year, the mill would require 53,000 hectares of short-fiber or 61,000 hectares of long-fiber growth to provide the required feedstock. The base case estimates of PNVs for tropical and subtropical locations using a 5 percent discount rate range from $1,828 per hectare for short fiber in West Africa to $3,649 per hectare for pine in Chile (see table 5-1). Applying these estimated PNVs to the land area requirements gives estimates of the amounts of development costs that could economically be incurred to establish an operating plantation. For short-fiber production in West Africa the maximum plantation acquisition and development costs would be almost $100 million. For the long-fiber production on a Chilean plantation the maximum amount is about $220 million.

By comparison, the acquisition and development costs that can be incurred by the investor in the slower-growing, lower-PNV temperate zones are much less. Since the average plantation yields are lower, the size of the plantation required to service a 260,000-metric-ton-per-year bleached kraft pulp mill will be larger. Assuming a plantation yield of 10 cubic meters per hectare per year, the land area required for the long-fiber-producing plantation would be about 122,000 hectares for the 260,000-metric-ton mill.

It should be noted that the investor may choose to develop a greater land area as insurance against low yields, fire, infestation, and so forth. However, unless any additional output in excess of that required by the mill can be successfully marketed, the net returns are unchanged, and hence, the fixed amount allowable for development costs would simply need to be spread over a greater land area.

In addition, of course, some investor must provide the $300 million required for the pulp mill and power plant complex. Hence, the total initial investment in mill, power plant, and acquisition and development costs (exclusive of site preparation and subsequent management activities) is as high as $400 million to $520 million.

Development Costs by Region

The examples above demonstrate that although very substantial development costs might be incurred in the establishment of forest plantations, the plantations can still generate substantial economic returns, provided that the other cost, yield, and return assumptions are realized. Table 4-1 presents a qualitative assessment of forest plantation development costs in the regions under consideration.

Our judgment is that in the developed countries, development costs will generally be relatively low. To a large extent this reflects the fact that much of the necessary infrastructure is already in place. For much of Brazil the assessment is for low-medium development costs, although in many cases, particularly where accessible former croplands are used, the development costs here are likely to be low. For Chile, the development costs are judged to be low, since plantations are located largely on formerly marginal agricultural lands with reasonably good access. Borneo is also assessed as medium, since existing logging operations in the native forest have provided some of the required infrastructure in the form of roads and so forth. Finally, the high development cost category is reserved for Amazonia and West Africa, the least economically developed of the regions under investigation.

TABLE 4-1. Estimated Assessment (High, Medium, or Low) of Forest Plantation Development Costs

Forest-producing region	Assessment
Traditional	
Nordic	Low
Pacific Northwest	Low
U.S. South	Low
Nontraditional	
Australia	Low
Borneo	Medium
Brazil	
Amazonia	High
Central Brazil	Low-Medium
Southern Brazil	Low-Medium
Chile	Low
New Zealand	Low
South Africa	Low
West Africa	High

Considerations Involving Risk

Besides development costs and acquisition costs, some provision must be made in the analysis for risk. Given the long-term nature of the investment commitment, there is some probability of the occurrence of biological or political events that could seriously reduce the economic returns from the investment.

One way of adjusting for risk is to incorporate a risk factor into the discount rate. If the riskless real discount rate was 5 percent, a political risk premium could be added to the riskless discount rate. For example, an appropriate risk premium for a country having high political risk might be 10 percent. Therefore, the discount rate to use in the PNV calculations would be 15 percent (5 percent real plus 10 percent for risk).

Economic Implications of Biological Risks

As with most economic activities, forest plantations have associated with their development a certain set of risks such as those involving control of costs, reasonable estimates of yields, and prices available in future markets. In addition, forest plantations are subject to uncertainties regarding the nature and magnitude of biological risks. In temperate forest plantations that utilize indigenous species, a large body of knowledge is typically available about the incidence and severity of disease, pests, insect infestation, fire, and so forth. However, when introducing exotic species into tropical regions, similar information is typically very sparse. (See chapter 8 for detailed discussion of the ecological implications of tropical plantation forestry.)

The question of whether exotic (nonindigenous) forest plantations are indefinitely sustainable in certain tropical regions remains unresolved (Fearnside and Rankin, 1980; Johnson, 1976). The question of the level of soil nutrients, particularly as the plantation moves beyond the first cycle of plantation cropping, is important to the long-run viability of the plantation in both biological and economic terms. Removal of the original forest (where this has occurred) and subsequent periodic removal of large amounts of biomass will result in the decline of the total stock of nutrients unless they are replaced. The geologically old soils of certain tropical regions cannot supply the mineral nutrients through weathering of igneous or young rocks. Fortunately the vast majority of nutrients are found in the leaves (needles) and branches of the trees which, if left at a site, are likely to be recycled into the soil. However, on fragile soils the nutrient losses from the removal of the stems (logs) could become important. To some extent, these losses can be offset through fertilization; however, the costs of fertilization will, of course, affect the economic returns of the plantation. While the overall question of the extent and rate of degradation of the soil and the ability and costs of offsetting this degradation remain unresolved for some sites and regions, Sanchez and coauthors (1982) provide a basis for optimism.

The possibility of pests and diseases is always a danger, particularly for a monoculture of an exotic species. This is especially true in moist tropics where the low density of individuals of the same species provides a defense against large-scale forest destruction. The substitution of an exotic monoculture that may be susceptible to indigenous pests or diseases therefore poses potential risks. The decimation of the rubber plantations at Fordlandia in the Amazon between 1926 and 1945 is a classic example.

Possible environmental impacts that may occur are of two types: (1) those that directly affect the plantation and plantation lands, and (2) those that affect areas beyond the plantation. The latter impacts could be relatively localized or global in nature. If the environmental effects are localized to the plantation—for example, depletion of nutrients in the soil, which reduces yields or requires cost-increasing offsets—the effect will be internalized to the firm and the costs (or expected costs) of these effects would be built into the decision function of a rational manager.

If the environmental effects move beyond the plantation, such as downstream pollution or flooding, they may be characterized as economic externalities. From a social point of view difficulties arise, since some of the costs are borne by a group other than the beneficiaries of the plantation activities. For the purpose of this study, however, the externalities associated with any particular plantation are assumed to be small enough to be ignored.

A final consideration relates to the role of the forest plantations and their impact upon the environment. While reforestation has generally been popular with environmental groups in industrial countries, the concept of the forest plantation, particularly when it replaces the natural forest, is disagreeable to many environmentalists. From an ecological point of view, however, the plantation forest is probably a better use of much of the land than most alternative commercial uses, such as croplands or pasture. In addition, an argument can be made that high-yield plantations potentially offer one of the better means of protecting the world's remaining natural forests, and particularly tropical, moist forests, from future destruction. To the extent that forest plantations can provide cheap industrial wood at locations accessible to major world markets, the logging of natural forests in increasingly inaccessible regions of the world becomes a dubious economic venture. This, however, does not guarantee the survival of the natural forests, since commercial logging is only one of a number of forces contributing to the pressures on natural forests.

Economic Implications of Political Risks

Political risk is mentioned in general terms here as a factor that could reduce economic returns flowing to the investor. Although losses to a plantation firm could certainly result from war and other international disputes involving more than one country, political risk usually would arise from the actions of or changes in the national government. For example, in regions of political instability, a government friendly to international investors might be replaced by a hostile one. Adverse governmental actions could run the gamut from outright expropriation of investor assets to either high or discriminatory taxes or both, stifling bureaucratic procedures, inconsistent and changing investment laws, and so forth. These risks are of particular importance for the private and foreign investor because forest plantation investments have a long gestation period (the shortest rotations are about six years, and multiple rotations are likely to be required to generate an adequate return on the investment, particularly if land acquisition and development costs are involved), and such plantations are typically associated with large investments in plants and equipment. Thus, the investor must be concerned about both the size of the investment and the long-time period during which it is vulnerable. The common practice of many foreign private investors undertaking investments in developing countries is to require that the expected period for the payback of all their investment capital be a very few years, perhaps three to five, reflecting the high discount rate they apply to investments in those countries to account for the largely political risk.

It should be noted, however, that the extent of this type of risk may vary with the type of investor. For example, a foreign private investor may run a greater risk of expropriation for political reasons than would a local private investor. The political risk is still less for a local public firm that is likely to survive political changes. Finally, the risk to the public owner may be much less than that to a private owner, since the asset is likely to remain socially productive even if the government changes.

Comparative and Absolute Advantage

Since a major objective of this study is to provide the reader with a sense of the potential of the various regions to compete in world markets as industrial wood growers, the concepts of absolute advantage and comparative advantage must be examined. In the present context, absolute advantage relates to the timber-producing and relative performance of the various regions. Comparative advantage relates not only to the ability of land within a region to produce timber, but also to its productivity for alternative economic activities, such as agriculture. Thus comparative advantage deals with the "best" economic activity in which land can be engaged, that is, the use that offers the highest economic return—including nontimber uses.

Suppose, for example, that a region such as Brazil grows timber and receives the highest economic return worldwide. Suppose further that those same lands might be able to generate even greater economic returns from the production of some other commodity, such as soybeans. If even higher economic returns to soybeans are obtainable elsewhere, Brazil would have an absolute advantage in the production of timber, while its comparative advantage would be in the production of soybeans.

The strict formal application of the model employed in this study provides a measure of the economic returns associated with a region's ability to grow timber (absolute advantage). The formal model together with other information provides insights into comparative as well as absolute advantage.

The relevant "other" information introduced into the broad study is derived primarily from actual experience about the types of investment activities being undertaken in market environments. In many of the twelve regions examined, markets are relatively unrestricted. For example, in the U.S. South and in Nordic countries, alternative uses of the land are possible. In fact, in the South and to a lesser extent in the Nordic countries, forests—commonly plantation forests—have recently displaced agriculture from significant portions of the land. This has occurred within the environment of a fairly unrestricted market. Similar experience is found in other regions. In New Zealand, marginal sheep pasture is now being converted to forest plantation because of the higher economic returns associated with tree growing. This is occurring despite public outcries lamenting the decline of traditional sheep raising. Likewise in Chile, the lands moving into industrial plantations often had been used for subsistence agriculture, which is the major economic alternative to forestry.

If markets are completely efficient, the market price of the land provides most of the information needed to determine its highest-value use. As noted earlier, in a perfect market the market price of the land represents the present net value of the land for its highest-value use. If plantation interests are willing to bid land away from a current nonforestry use at the market price, this is prima facie evidence that the forest plantation use is the land's highest-value use. Or expressed another way, forest interests will be willing to pay the market price generated by nonforestry uses only if the PNV in forestry is high enough to cover development costs plus pay the non-forestry-determined market price. Under these conditions the market price will come to reflect the forestry use, which is now the highest-value use. If the returns from the forest plantation activities are not sufficient to justify bidding the land away from the alternative use, given efficient markets, this simply reflects the fact that a nonforestry alternative use has a higher value than forestry has.

A second, related point should be raised. The current mix of land in croplands, pasture, and forest lands reflects an adjustment to the relative prices of the outputs of these activities. A rising real price of stumpage increases the return to forestry investments. However, the same situation holds for alternative uses. Should agricultural prices rise relative to forest prices, the relative returns will tip the balance in favor of cropland uses over forestry uses. Thus, without any changes in our formal analysis, the price of land for agricultural uses could rise relative to forest land, and an associated shift from forest to cropland would be expected.

In the regions of the tropics and temperate Southern Hemisphere, markets are sometimes less perfect than those in traditional forest-producing areas, and are more likely to be influenced by various types of governmental intervention. Nevertheless, for most of these nontraditional forest-producing regions, the evidence suggests that timber growing is the highest-value land use, and thus has a comparative advantage.

For example, in non-Amazonian Brazil, much of the plantation afforestation effort is occurring on lands that had been utilized for agriculture and then were abandoned due to declining soil fertility. While the afforestation efforts are subsidized, the fact that the land often had been lying idle suggests that the alternative commercial uses are few and that the returns to other investments would be low. Likewise, in the Amazon the alternative economic uses of the land appear to be limited, and generally the land is not involved in major commercial pursuits. For example, the Brazilian Jari plantation had previously been used for gathering ground nuts, and the entire area of over 1 million hectares was reputedly purchased for only $30 million, or less than $30 per hectare. This low purchase price can be interpreted as reflecting the low value of the alternative economic activities.

Similar situations exist in many of the other regions of the globe where industrial forest plantation activity is being undertaken. Almost without exception the plantations are being established on lands not used for major economic pursuits, or used only for marginal agricultural activity. In fact, the data used in this study are drawn very largely from the experience of forest plantations that have displaced other land uses or use previously idle lands.

Given this situation, the use of the land for forest plantations, together with the estimates of relatively high PNVs for forest plantations, suggests that plantations are the highest-value use of the land, that is, the use in which the land has a comparative advantage.

Alternative Processing Structures

The basic model used for this study assumes that the wood resource is processed in the region where it was grown. The basis for this assumption is simply that the transportation of wood in processed form generates the least cost output. The per unit volume costs of transporting wood as woodchips are about twice the per unit volume costs of transporting wood in the form of wood pulp. Therefore, it will not generally be economically rational over the long run to transport wood chips between regions on a large scale. However, some exceptions may be found. Obviously, in a case where local processing facilities either do not exist or are insufficient to exhaust the total available fiber resource, the local fiber price may be bid low enough to allow for its large-scale exportation to other regions that have sufficient processing facilities. The large flows of woodchips from the Pacific Northwest region to Japan may reflect such a situation. This situation might also occur if local processing costs are abnormally high. As will be seen in the sensitivity analysis presented in chapter 6, the effect of higher wood pulp processing costs on stumpage prices would be to depress stumpage prices and ultimately the economic returns to forest plantations located in regions with abnormally high processing costs. Under such conditions it may be economically rational to rely upon processing facilities outside of a region.

For illustrative purposes the following questions are examined: Under what conditions might Nordic processers find it economic to import woodchips from southern Brazil, and what type of economic return would this generate to the forest plantation investments of southern Brazil? The same questions could be addressed for any other regions examined in this study. For this exercise it is assumed, as throughout this study, that no artificial trade barriers exist. Given that the transport costs of a unit volume of wood shipped as woodchips are twice the transport costs of wood shipped as wood pulp, it is estimated that the Nordic countries could afford to import woodchips if stumpage prices in southern Brazil were roughly 75 percent or less of stumpage prices for the comparable resource in the Nordic region. Then the Nordic region could incur the additional transport costs associated with the Brazilian woodchips and still have the delivered Brazilian wood at a price comparable with that of domestic wood. Abstracting from legitimate concerns about the impact upon domestic prices and the stability of the wood supply source, wood processors would be indifferent to using domestic or foreign wood.

What are the implications of this scenario for the forest plantation investments in southern Brazil? A stumpage price of 75 percent of the Nordic base case gives a stumpage price for southern Brazil of $15.74 per cubic meter. For this stumpage price the PNV with a 5 percent discount rate is $2,830 and the IRR is $14.04. Thus, using our economic criteria the returns to forest plantation investments in southern Brazil, abstracting from both development costs and risk, appear substantial. More broadly, the analysis suggests that in some regions the potential for industrial forest plantations is independent of the economic viability of wood processing in those regions.

Exchange Rates

The relative costs and prices that are built into this study reflect the situation that existed circa 1979. A comparison of costs in different countries requires the use of an exchange rate whereby local currency costs are converted into a common currency, in this case the U.S. dollar. Some difficulties arise, however, in obtaining accurate comparisons. First, the domestic price level may change in some of the countries under consideration. If the exchange rate were freely floating, an offsetting adjustment would occur. Often, however, the exchange rate is not allowed to make the adjustment. Thus, for example, high inflation rates in Brazil that were not offset by an exchange rate depreciation would imply a corresponding increase in the dollar costs in Brazil that would be in excess of those that would be experienced in the other regions. The effect of the inflation would be to increase Brazil's cost structure relative to the other countries under consideration. This would adversely affect the relative economic position of Brazilian forest plantations and would be reflected in a reduced dollar value of their PNV. It should be stressed, however, that the effect of domestic inflation on the dollar prices of Brazilian expenses could be fully offset by an exchange rate devaluation just equal to the domestic price inflation. Furthermore, countries with high rates of domestic inflation relative to the worldwide norm commonly allow, or are forced by market pressures to allow, their exchange rate to depreciate to facilitate an offset.

The second type of exchange rate adjustment that could disrupt the relative relationship of 1979 would be an exchange rate change in the absence of offsetting domestic price-level changes. Either appreciating or depreciating the exchange rate can cause all of a country's domestic prices, when converted into dollar terms, to rise or fall and thereby increase or decrease the country's comparative cost structure. In our study such an occurrence again would be reflected by a change in the PNV.

A third reason for an exchange rate adjustment would be to reflect the increasingly (or decreasingly) competitive nature of a country's productive capacity. Exchange rate changes of this type can be expected to affect marginal domestic economic activities. Suppose, by way of example, that Brazil, a large and extremely energy-dependent country, suddenly found vast reserves of petroleum. The long-term effect of the new domestic energy resources would be likely to improve substantially the Brazilian balance of payments and to create pressures for an appreciation of the Brazilian currency. The effect would be to raise Brazilian costs in dollar terms and ultimately to reduce the competitive position of Brazilian forest plantations. Such a change does not necessarily imply that forest plantations would cease to be economic. However, the PNV of forest plantations would decline and some super marginal plantations would likely become submarginal.

Although considerations of long-run exchange rate changes are important, it is beyond the scope of this study to attempt to predict such changes systematically. Implicitly, the study assumes that the exchange rate structure will remain unchanged. Although this is certainly unlikely, it is probably better than any single alternative assumption that might be made.

References

Fearnside, Philip M., and Judy M. Rankin. 1980. "Jari and Development in the Brazilian Amazon," *Interciencia* vol. 5, no. 3, pp. 146–156.

Johnson, Norman E. 1976. "Biological Opportunities and Fast-Growing Plantations," *Journal of Forestry* vol. 74, no. 4, pp. 206–211.

Sanchez, Pedro A., Dale E. Bandy, J. Hugo Villachica, and John J. Nicholaides. 1982. "Amazon Basin Soils: Management for Continuous Crop Production," *Science* vol. 216, pp. 821–827.

5

Model Results: The Base Case

The base case is used as the benchmark for the analysis. This case reflects 1979 costs, prices, yields, and so forth, extended in perpetuity. The base case incorporates the simplifying assumption that processing and harvest costs are essentially the same for all the regions examined. This allows the analysis to focus on the economics of timber growing. While the base case provides the standard to which all of the sensitivity analysis scenarios are compared (see chapter 6), it should be recognized that for some of the regional plantations, selected alternative (sensitivity analysis) scenarios might provide a better assessment of the existing or likely future economic situation than does the base case.

The base case, presented in tables 5-1, 5-2, and 5-3, examines both pulpwood and integrated (sawtimber) regimes for each representative plantation. In all cases short-fiber[1] final products, whether wood pulp or raw lumber, are assumed to sell at a price 10 percent below that of comparable long-fiber

[1] In this study gmelina and eucalyptus are short-fiber species.

products. In addition, one variant of the base case (column 3 of the same tables) treats lumber from nontraditional sources, that is, sources other than North America or Europe, as being of lower quality and thus discounted 10 percent in final markets. This variant reflects a common perception that tropical and temperate Southern Hemisphere softwood lumber is of lower quality than is its Northern Hemisphere counterpart.

Base Case Findings

The base case results for present net value (PNV) are found in tables 5-1 (5 percent discount rate) and 5-2 (10 percent discount rate). Table 5-1 reports the estimated present net values for both a pulpwood regime and an integrated (sawtimber/pulpwood) regime for each representative plantation. As explained earlier, the calculated values are based upon the costs, management, and yields of the representative plantations, together with the implicit stumpage prices associated with the dominant world submarket, that is,

TABLE 5-1. Representative Plantations, Base Case; Present Net Value, 5% Discount Rate; 1979 Constant Prices in Perpetuity
(1979 U.S. $/hectare)

	Regime		
Region/species	Pulpwood	Integrated, with standard-quality sawtimber	Integrated, with lower-quality sawtimber[a]
North America			
U.S. South			
Pinus taeda, avg.-yield site	1,748	2,474	—
Pinus taeda, high-yield site	2,830	3,742	—
Pacific Northwest			
Pseudotsuga menziesii, avg.-yield site	616	902	—
Pseudotsuga menziesii, high-yield site	1,077	2,248	—
South America			
Brazil, Amazonia			
Pinus caribaea	3,087	4,080	3,376
Gmelina spp.	2,530	3,184	—
Brazil, Central			
Eucalyptus spp.	3,456	4,027	—
Brazil, Southern			
Pinus taeda	3,592	4,715	3,953
Chile			
Pinus radiata	3,649	4,509	3,206
Oceania			
Australia			
Pinus radiata	2,005	2,141	1,378
New Zealand			
Pinus radiata	2,903	4,118	2,567
Africa			
South Africa			
Pinus patula	3,051	3,727	2,783
Gambia-Senegal			
Gmelina spp.	2,289	2,622	—
Eucalyptus spp.	1,828	3,262	—
Europe			
Nordic			
Picea abies	−100	154	—
Asia			
Borneo			
Pinus caribaea	1,851	2,364	1,846

[a] This column gives estimates of PNV for non-traditional producers where the sawtimber final product (lumber) is assumed to sell at a 10 percent discount when compared to the price of lumber from traditional suppliers.

the submarket that generates the highest implicit stumpage price.

In almost all cases presented in table 5-1 the PNV associated with the integrated regime (column 2) is higher than that for the pulpwood regime (column 1). This result suggests that for the regions examined, the integrated regime with standard-quality sawtimber is economically superior, given the assumptions of the analysis.

Column 3 in table 5-1 presents the results for the seven relevant regions when the price of softwood (lower-quality) lumber of the tropical and temperate Southern Hemisphere regions is reduced by 10 percent. When the lumber price is discounted 10 percent,

TABLE 5-2. Representative Plantations, Base Case; Present Net Value, 10% Discount Rate; 1979 Constant Prices in Perpetuity
(1979 U.S. $/hectare)

Region/species	Pulpwood	Regime Integrated, with standard-quality sawtimber	Integrated, with lower-quality sawtimber[a]
North America			
U.S. South			
Pinus taeda, avg.-yield site	190	276	—
Pinus taeda, high-yield site	452	555	—
Pacific Northwest			
Pseudotsuga menziesii, avg.-yield site	−278	−342	—
Pseudotsuga menziesii, high-yield site	−119	−55	—
South America			
Brazil, Amazonia			
Pinus caribaea	850	1,232	991
Gmelina spp.	944	1,052	—
Brazil, Central			
Eucalyptus spp.	1,143	928	—
Brazil, Southern			
Pinus taeda	865	1,262	1,035
Chile			
Pinus radiata	1,019	908	605
Oceania			
Australia			
Pinus radiata	99	10	−153
New Zealand			
Pinus radiata	344	652	201
Africa			
South Africa			
Pinus patula	847	844	592
Gambia-Senegal			
Gmelina spp.	682	693	—
Eucalyptus spp.	280	775	—
Europe			
Nordic			
Picea abies	−497	−417	—
Asia			
Borneo			
Pinus caribaea	302	502	345

[a] See table 5-1, footnote a.

the pulpwood regime for many of the regions generates a higher PNV than the integrated regime with lower-quality sawtimber does. Thus, this fairly small change in the world market lumber price will shift the economics of several of the regions away from an integrated regime toward a pulpwood regime.

It will also be observed that in the two cases—Amazonia/*Pinus caribaea*—where the PNV of the integrated regime with lower-quality sawtimber continues to exceed the PNV of the pulpwood regime, the difference is quite modest, and small additional reductions in lumber prices would surely tip the balance in favor of the pulpwood regime.

One implication of the above is that the

uncertainty as to the degree of acceptance of tropical and temperate Southern Hemisphere softwood lumber may not be as important as is sometimes believed, since in most cases fairly modest lumber price reductions will simply result in a shift to a wholly pulpwood-oriented plantation regime.

Of the PNVs presented in table 5-1 for thirty-two regimes (sixteen pulpwood and sixteen integrated with standard-quality sawtimber), all except one of the regimes have a positive PNV using a 5 percent discount rate. The regime with negative PNV at the 5 percent discount rate is the Nordic pulpwood regime.

Raising the discount rate to 10 percent (see table 5-2) generates a number of changes. Of course, the PNV of every region and regime will decline. The total range of PNVs at the 5 percent rate is −$100 (Nordic region) to $4,715 (southern Brazil). At a 10 percent discount rate, the range falls to −$497 (Nordic region) to $1,262 (southern Brazil). Also at the higher discount rate, seven plantation regimes generate negative PNVs. These are the pulpwood and integrated (standard-quality sawtimber) regimes of the Pacific Northwest, both average-yield and high-yield sites, and the Nordic region pulpwood and integrated regimes. In addition, the higher discount rate places a premium on short-payoff investments. Thus, at the 10 percent discount rate, five of the sixteen cases presented have the higher PNV associated with the shorter, pulpwood regime.

Table 5-3 presents an alternative investment criterion, the internal rate of return (IRR) for the plantations and regimes under examination. This criterion, in essence, uses its endogenously determined IRR as the discount rate. Therefore, plantations with high IRRs tend to favor short-rotation, faster-payoff pulpwood investments. This is reflected in the seven out of sixteen cases in which the pulpwood regime generates a higher IRR than does the integrated regime. The IRRs generated in the base case range from 4.61 percent for Nordic pulpwood to 27.53 percent for Amazonian hardwood (gmelina) pulpwood.

Regional Differences

Although a simple ranking of plantations by PNV or IRR would reveal some differences, both investment criteria provide a generally consistent assessment. The regions of the tropics and temperate Southern Hemisphere rank consistently high, the plantation of the Nordic region consistently low, with Pacific Northwest plantations also generally ranking relatively low. However, the high-yield Pacific Northwest plantations begin to approach some of the lower-ranked Southern Hemisphere sites. The U.S. South, both the average-yield and high-yield sites, generally perform comparably with the tropical and temperate Southern Hemisphere plantations, with the high-yield site on a par with the better tropical and temperate Southern Hemisphere sites.

As expected, the introduction of a reduced lumber price for nontraditional producers does diminish the returns from integrated plantations in these regions. In five of the seven regions examined, the pulpwood option's return is superior to that of the integrated plantation with lower-quality sawtimber. In these cases, it would be expected that the plantation would eventually shift to an exclusively pulpwood operation, as the sawmill facilities would not justify replacement. In essence, the pulpwood option provides an effective floor on the long-run return available to the plantation.

NORTH AMERICA. Four representative plantations are examined in North America (two in the U.S. South—an average-yield and a high-yield site—and two representative of the Douglas fir region of the Pacific Northwest, both an average-yield and high-yield site). The U.S. South performs better than the Pacific Northwest by either the PNV or the IRR criterion. Using a 5 percent discount rate (table 5-1), the U.S. South high-

TABLE 5-3. Representative Plantations, Base Case; Internal Rate of Return, 1979 Constant Prices in Perpetuity
(percent)

Region/species	Pulpwood	Regime Integrated, with standard-quality sawtimber	Integrated, with lower-quality sawtimber[a]
North America			
U.S. South			
Pinus taeda, avg.-yield site	12.02	12.45	—
Pinus taeda, high-yield site	13.91	14.12	—
Pacific Northwest			
Pseudotsuga menziesii, avg.-yield site	7.11	7.07	—
Pseudotsuga menziesii, high-yield site	8.85	9.62	—
South America			
Brazil, Amazonia			
Pinus caribaea	17.89	20.44	19.25
Gmelina spp.	27.53	23.54	—
Brazil, Central			
Eucalyptus spp.	20.16	15.54	—
Brazil, Southern			
Pinus taeda	15.57	17.53	16.79
Chile			
Pinus radiata	23.39	17.50	16.01
Oceania			
Australia			
Pinus radiata	10.68	10.06	8.94
New Zealand			
Pinus radiata	11.90	13.11	11.13
Africa			
South Africa			
Pinus patula	19.34	17.69	16.27
Gambia-Senegal			
Gmelina spp.	18.42	17.52	—
Eucalyptus spp.	12.16	14.71	—
Europe			
Nordic			
Picea abies	4.61	5.57	—
Asia			
Borneo			
Pinus caribaea	12.94	14.73	13.61

[a] See table 5-1, footnote a.

yield site plantations generate PNVs of close to $3,000 per hectare for the pulpwood regime and over $3,700 per hectare for the sawtimber regime. Using the IRR criterion (table 5-3), the U.S. South again performs well, with the IRRs being between 12 percent (for the average-yield site/pulpwood regime) and 14 percent (for the high-yield site/integrated regime).

The returns generated by the Pacific Northwest plantations are quite acceptable, but not outstanding. Using a 5 percent discount rate, the PNVs range from an acceptable $616 per hectare (average-yield site/

pulpwood regime) to a healthy $2,248 (high-yield site/integrated regime). The IRRs range from 7.07 percent to 9.62 percent.

SOUTH AMERICA. Many analyses view South America as having perhaps the greatest potential of all regions for the development of plantation forests. The results of this study bear out that optimism. By either the PNV or the IRR criterion, all the South American regions examined perform exceptionally well. Five representative plantations are investigated—two in Amazonia, one in central Brazil, one in southern Brazil, and one in Chile. All five regions compare favorably with any regions examined in this study, by either criterion. For Brazil, the best-performing region utilizing the PNV criterion is southern Brazil with $4,715, although both central Brazil and Amazonia also perform very well. The Chilean results are comparable with the best performers in all regions.

OCEANIA. New Zealand and Australia both also perform very well by either criterion. New Zealand in particular rivals the best performance of the Brazilian plantations using the PNV criterion ($4,118), and is just a step behind using the IRR criterion (13.11 percent). Introducing the two-quality lumber products and selling New Zealand and Australian lumber at a price discounted 10 percent reduces the PNV and IRR only modestly, but would cause a long-run shift to specialization wholly in pulpwood.

AFRICA. The African plantations all performed well. South Africa in particular generates returns by either criterion that slightly exceed the high-yield site for the U.S. South with a PNV of $3,727 and an IRR of 19.34 percent. Both West African plantations performed respectably. Reducing the South African lumber price 10 percent resulted in a significant but not overwhelming reduction in the returns. However, the reduction is sufficient to result in a shifting of production away from lumber and wholly into pulpwood.

EUROPE. The slow-growing Nordic forests are generally the poorest performing of all plantations examined. The PNVs shown in table 5-1 are below zero in the case of the pulpwood regime (−$100) and only slightly above for the integrated regime ($154). The IRRs are around 5 percent. This relatively poor performance persists despite the fact that the Nordic forests generate the highest stumpage prices, a reflection of their accessibility to the high-price northern European market.

ASIA. Only one region of Asia—Borneo in the East Indies Archipelago—was investigated. It should be noted that the data used for this region are probably the least reliable of all used in this study. Given this caveat, the performance of this region was quite acceptable, PNV being $1,851 for the pulpwood and $2,364 for the integrated regimes. These results are roughly comparable by either criterion to the average-yield site in the U.S. South.

Summary and Interpretation of Findings

Table 5-4 summarizes the number of positive and negative PNV estimates for each regime for discount rates of 5 and 10 percent. Using a 5 percent discount rate, thirty-eight of the thirty-nine regimes generated positive PNVs. For the 10 percent discount rate there were positive PNVs for thirty-two of thirty-nine regimes. Thus the overwhelming majority of the plantations passed the feasibility criteria based on plantation management cost, stumpage price estimates, and biological considerations.

However, at this point the analysis is not complete, and the PNVs and IRRs must be interpreted with care. Values of these common investment criteria presented here re-

TABLE 5-4. Summary of Base Case Positive and Negative Results

Regime	Positive	Negative
5 percent discount rate		
Pulpwood	15	1
Integrated, with standard-quality sawtimber	16	0
Integrated, with lower-quality sawtimber	7	0
Total	38	1
10 percent discount rate		
Pulpwood	13	3
Integrated, with standard-quality sawtimber	13	3
Integrated, with lower-quality sawtimber	6	1
Total	32	7

flect the economics of investments in forest management activities alone and do not consider other costs, for example, acquisition and development costs and risk, which are discussed in the preceding chapter. For example, as stated earlier, development costs refer to the costs associated with the introduction of the necessary infrastructure, physical and social (if necessary), required to allow for undertaking the set of activities characterized as a management regime. If development costs (as well as acquisition costs and those associated with biological or political risk) are either nonexistent or if they could be expected to be of similar magnitudes for all the plantations, then a simple ranking by PNV would reflect totally the comparability of the representative plantations. However, the study compares plantations that are representative of regions in very different states of development. For the traditional forest-producing countries and also probably in Australia, New Zealand, South Africa, and the other developed countries, development costs are likely to be small since well-developed infrastructures generally exist. Therefore, for these regions the forest management costs are generally representative of the true economic costs associated with the forest plantation investment. However, for some of the nontraditional producing regions, development costs associated with the establishment of a forest plantation may be quite large, and therefore must be included in a complete evaluation.

Without quantifying development costs, land acquisition costs, and risk, the formal model provides information necessary to a comprehensive assessment of economic viability of development investments in forest plantations. The estimate of present net value for each representative plantation provides an estimate of the capitalized value of a regionally representative forest site for timber production over a perpetual time horizon. An economic interpretation of the meaning of the various PNVs is that it indicates the amount of initial investment that the rational investor—or, more broadly, society—should be willing to pay in order to make that land available for management investments in forest plantation operations of the type discussed above. In other words, the PNV is the price the investor should be willing to pay for property rights to develop and use the forest plantation site. If the acquistion and development costs are equal to the PNV, the implication is that the project will return to society a percentage return equal to the discount rate used in the analysis. If the acquisition and development costs are less than the PNV, the return will be greater than the discount rate used in the analysis. If the acquisition and development costs are greater than the PNV, the project returns do not cover the opportunity costs of the land and capital required to get the plantation on stream.

Sources of Differential Performance

This section examines the sources of the differences in the present net values that are presented above. The examination here is nonrigorous, and considers individually the effects of rotation lengths, biological yields, stand establishment costs, and subsequent silvicultural costs, upon calculated PNV. The section demonstrates that the economic viability of the forest plantations, as measured in this study by the investment criteria, is dependent upon the interaction of all the above-mentioned factors—rotation lengths, yields, establishment cost, and subsequent costs—as well as stumpage prices. It also shows that no single factor explains the relative economic viability of the representative plantations. However, short rotations and high biological yields, as well as a high stumpage price, are important to a successful performance using the investment criteria.

Effect of Rotation Lengths, Yields, and Stand Costs on PNV

It is clear that stumpage values alone do not determine the relative economic viability of the various regions examined. For example, among the twelve regions, the Nordic region has the highest stumpage prices for the base case and for virtually every sensitivity analysis scenario (stumpage prices for base case and sensitivity analysis scenarios are given in tables D-1 through D-7 in appendix D). However, the Nordic region's performance in terms of the investment criteria (tables 5-1, 5-2, and 5-3) was almost always the poorest of the regions examined. Thus other factors—specifically rotation lengths, biological yields, stand establishment costs, and subsequent silvicultural costs—have a major influence upon a plantation's investment performance.

It is well known that with a positive interest rate, current returns are preferred to future returns and, conversely, future costs are preferred to current costs. This consideration alone suggests an economic advantage for investment projects with short payoff periods over projects with long payoff periods, *ceteris paribus*. Therefore, short-rotation plantations have an apparent economic advantage over longer-rotation plantations.

However, not only do rotations vary, but so do quality of output and biological yields.

In its construction of the representative plantations and in its various calculations, this study takes into account both different outputs—pulpwood, sawtimber, hardwood, and softwood, with different associated values—and also different biological yields. Different yields will, of course, generate different receipt levels and therefore, other things being equal, higher yields are preferable to lower yields in terms of their effect on the investment criteria.

Costs obviously affect the economic viability of any investment. For plantations, development costs vary considerably, as discussed in detail in chapter 4, depending upon the location of the plantation, the types of infrastructure, and so on, that are already in place. The second type of costs is establishment costs incurred in the first year of the rotation in the site preparation and planting associated with stand establishment. The third, separate set of costs, referred to as subsequent costs, are those that occur subsequent to plantation establishment. Table 5-5 presents the initial establishment (site preparation and planting) costs and subsequent costs for each representative plantation.

Initial establishment costs have a close relationship with PNV estimates obtained. Since the present net value represents the differences between discounted benefits (receipts) and discounted costs (expenditures) in the initial year of the investment, any increase in initial-year establishment costs will simply be added to the discounted costs and thereby reduce the PNV accordingly, that is, a dollar added to initial-year establishment costs will reduce the PNV by a corresponding dollar, and a more complex sensitivity analysis is unnecessary.

The effect of subsequent costs is similar to that of establishment costs except that since they are incurred in later years the cost must be discounted back to the initial period. Therefore, a $1 increase in subsequent costs will reduce PNV by something less than $1, depending upon the point in time in which the cost is incurred and the discount rate that is used.

TABLE 5-5. Representative Plantations: Undiscounted Initial and Subsequent Costs of Stand Establishment
(1979 U.S. $/hectare)

Region/species	Pulpwood regime		Integrated regime	
	Initial costs	Subsequent costs	Initial costs	Subsequent costs
North America				
U.S. South				
Pinus taeda, avg.-yield site	266.17	139.15	266.17	149.15
Pinus taeda, high-yield site	266.17	139.15	266.17	149.15
Pacific Northwest				
Pseudotsuga menziesii, avg.-yield site	327.64	395.29	327.64	712.92
Pseudotsuga menziesii, high-yield site	327.64	370.29	327.64	687.92
South America				
Brazil, Amazonia				
Pinus caribaea	277.50	370.00	277.50	430.00
Gmelina spp.[a]	145.00	810.00	145.00	375.00
Brazil, Central				
Eucalyptus spp.[a]	523.00	647.80	523.00	647.80
Brazil, Southern				
Pinus taeda	523.00	382.80	523.00	502.80
Chile				
Pinus radiata	176.62	92.47	176.62	114.38
Oceania				
Australia				
Pinus radiata	651.56	173.73	651.56	192.51
New Zealand				
Pinus radiata	456.33	445.46	456.33	567.40
Africa				
South Africa				
Pinus patula	184.69	195.00	184.69	318.30
Gambia-Senegal				
Gmelina spp.[a,b]	350.00	250.00	450.00	265.00
Eucalyptus spp.[a,c]	961.00	240.00	961.00	355.00
Europe				
Nordic				
Picea abies	456.00	126.72	456.64	156.72
Asia				
Borneo				
Pinos caribaea	277.50	415.00	277.50	490.00

Note: Initial costs = those for site preparation and planting associated with stand establishment. Subsequent costs = those occurring after stand establishment.
[a] Three harvests per planting (root stock) in pulpwood regime.
[b] Two harvests per planting (root stock) in integrated regime.
[c] Three harvests per planting (root stock) in integrated regime.

Study Experience with Rotations, Yields, and Harvests

Table 5-6 summarizes the experience and practices of the twelve study regions with respect to rotations and harvest volumes. The rotations vary in length from six years for the second and the third gmelina and eucalyptus pulpwood rotations in Amazonia and central Brazil to eighty years for a sawtimber rotation in the Nordic region (integrated regime). The two six-year rotations involve multiple harvesting from a given root system based upon coppicing (forest growth origi-

TABLE 5-6. Representative Plantations: Rotation, Growth, and Yields
(prices in 1979 dollars)

Region/species	Pulpwood regime			Integrated (sawtimber) regime		
	Rotation age (yrs)	Mean annual increment m³/ha/yr	Total commercial yield ($/ha)	Rotation age (yrs)	Mean annual increment m³/ha/yr	Total commercial yield ($/ha)
North American						
U.S. South						
Pinus taeda, avg.-yield site	30	11.90	356.67	35	12.40	433.93
Pinus taeda, high-yield site	39	17.93	538.02	35	17.72	620.20
Pacific Northwest						
Pseudotsuga menziesii, avg.-yield site	40	12.75	510.00	50	14.42	721.00
Pseudotsuga menziesii, high-yield site	30	13.60	408.00	40	20.03	801.00
South American						
Brazil, Amazonia						
Pinus caribaea	12	16.00	192.00	16	16.00	256.00
Gmelina spp.	7[a],13,18	18.00	342.00	12	18.00	216.00
Brazil, Central						
Eucalyptus spp.	7[a],13,18	25.00	475.00	19	25.00	475.00
Brazil, Southern						
Pinus taeda	12	20.00	240.00	20	20.00	400.00
Chile						
Pinus radiata	25	22.00	555.00	32	21.98	703.30
Oceania						
Australia						
Pinus radiata	29	20.69	600.00	35	17.40	609.00
New Zealand						
Pinus radiata	18	25.00	450.00	27	25.00	675.00
Africa						
South Africa						
Pinus patula	15	16.10	241.50	26	16.00	416.00
Gambia-Senegal						
Gmelina spp.	10[a]	15.00	450.00	20,40	15.00	600.00
Eucalyptus spp.	7[a]	17.00	357.00	10,20,30	17.00	510.00
Europe						
Nordic						
Picea abies	50	5.00	250.00	80	5.00	400.00
Asia						
Borneo						
Pinus caribaea	15	14.00	210.00	20	13.50	270.00

[a] Three harvests per root stock.

nating mainly from shoots rather than seeds) after the first harvest. The average annual yield in cubic meters per year varies from a low of 5.0 in the Nordic region to a high of 25.0 for eucalyptus in central Brazil and pine in New Zealand.

For base case as well as for most of the scenarios examined in the sensitivity analysis in chapter 6, the short-rotation plantations perform better under the economic criteria than do the very long rotations. This is especially true for the IRR criterion that implicitly uses, for most representative plantations examined, a higher (and therefore short-term biased) discount rate. It should be noted, however, that some of the better performers have intermediate rotation lengths. These include Chile, New Zealand, and the U.S. South. Thus, while rotation length is positively related with economic performance, other factors also influence the final economic performance.

Biological yield is also a factor to be considered when analyzing the elements that contributed to a strong performance using the investment criteria. However, some plantations with lower yields such as in Amazonia perform better by the economic criteria than do the higher-yield plantations, for example, southern Brazil and Chile. In addition, while the Pacific Northwest high-yield site integrated plantation has a relatively high yield of 20.03 cubic meters per hectare per year (table 5-6), the economic performance of the PNW plantation as reflected in the investment criteria is clearly poorer than that of several other representative plantations (such as the U.S. South) with lower yields.

The lowest initial establishment costs, as shown in table 5-5, are found in South Africa ($184.69/hectare), Chile ($176.62/hectare) and the gmelina plantation of Amazonia ($145.00/hectare). The highest are those of the Gambia-Senegal eucalyptus establishment ($961.00/hectare) and the eucalyptus plantation of central Brazil ($523.00/hectare). However, these are the highest establishment costs for the three-cycle crops that are coppicing species, and hence do not provide a direct comparison with the establishment costs for more typical one-cycle crops. For the one-cycle plantations the high-cost regions are Australia ($651.56/hectare), and southern Brazil ($523.00/hectare). Costs subsequent to establishment also vary considerably, with the lowest undiscounted costs being experienced in the Nordic region, the U.S. South, Australia, and Chile. High subsequent undiscounted costs are experienced in the Pacific Northwest, New Zealand, and Brazil. (See table 5-5 for subsequent costs for pulpwood and integrated regimes.)

There is no clear and simple relationship between establishment and subsequent costs. High-establishment-cost regions are found among the strongest performers using the economic criteria. For example, most Brazilian plantations have relatively high establishment costs. A similar lack of a clear relationship with PNV also applies to undiscounted subsequent costs.

Economic Performance of Selected Representative Plantations

It is helpful to relate the above-mentioned factors to the relative performance of the investment criteria of selected representative plantations. The high-yield-site PNW plantation, for example, has as a return-enhancing feature in terms of relatively high biological yields. However, the PNW also has long rotations and high establishment and subsequent costs, as well as low stumpage prices. Overall, although the PNW performs acceptably, its performance by investment criteria is not as strong as that of some other regions. The U.S. South, by contrast, has generally similar biological yields. However, the U.S. South, with shorter rotations, lower establishment and subsequent costs, and higher stumpage prices, performs significantly better using the investment criteria. The central and southern Brazil and Amazonian plantations have relatively short rotations, high stumpage prices (except the hardwood, which is somewhat depressed because of the discounted price of its final product in world markets), and high biological yields. These features are sufficient to generate high performance according to the investment criteria despite the rather substantial establishment and subsequent costs generally experienced. New Zealand also performs well by the investment criteria, reflecting the positive effects of relatively low establishment and subsequent costs and rapid biological growth, and in spite of the somewhat low stumpage prices and a long (by tropical standards) rotation. Finally, the Nordic forest plantation performs least well using the investment criteria, reflecting a long rotation, low biological growth, and fairly high establishment costs. Only the high stumpage prices for the Nordic plantations are a force for improving its relative performance using the investment criteria.

6

Sensitivity Analysis

In the previous chapter investment criteria were used to assess the economic returns to the regionally representative plantations for the base case. As in all models of this type, the results, that is, the estimates of the present net values (PNVs) and the internal rates of return (IRRs), are a reflection of the cost, volume, price, and other data that are introduced into the model. However, all of these data are subject to error, and even if they perfectly reflected the values as of 1979, changes in the values would undoubtedly occur over the extended time period covered by the analysis. Because of the difficulties of obtaining detailed data for certain regions, as explained in chapter 2, the study has assumed constant costs across regions for some production activities (namely, harvesting, internal transport, and processing), although as noted earlier, certain geographic regions may have inherently higher or lower costs for those activities. To make allowance for these limitations, this chapter examines the effects of introducing separate and independent changes in the costs and prices used for the base case estimates of PNVs and IRRs.

Such an analysis provides the researcher and interested reader with a notion of the sensitivity of the PNV estimates to changes in the magnitude of costs and prices.

A scenario approach also allows the reader to make additional comparisons, incorporating information that may not be widely available. For example, an individual may believe that the assumptions of the model are basically sound except that wood pulp processing costs are likely to be 20 percent higher in the tropics than in the Northern Hemisphere temperate climate regions. Meaningful comparison then would encompass a contrast of the base case investment criterion results for the temperate regions with the results of the high wood-pulp-processing-cost scenario for the tropical regions.

Let us review the premises of the base case before taking up the sensitivity analyses. The base case of this study incorporates a set of costs, yields, and prices that applied circa 1979. The management costs and biological yields are derived from regionally specific data that are believed to be representative values for these variables. The stumpage prices used for each region are shadow prices derived from prices in final markets and from the costs involved in harvesting the timber, processing into wood pulp or lumber, and transporting the final product from the region of processing to the world market. The harvesting and processing costs are assumed to be identical for all plantations worldwide. The final important set of costs—international transport costs—are estimated on the basis of distance. All prices are converted to 1979 U.S. dollars using official exchange rates.

Why Sensitivity Analysis is Required

As the preceding section explains, considering data limitations and changes that can be expected over time, sensitivity analysis is needed to provide flexibility in utilizing the model results. This section discusses in detail the considerations that form the background of the various sensitivity analyses.

Reasons for Differences Between Actual and Model Costs

There are a number of reasons why actual costs may differ from those used in the model. First, there are obviously site-specific considerations, and, as noted earlier, it is beyond the scope of this study to try to incorporate site-specific (as opposed to regionwide) costs. Second, there may be systematic cost differences by region. An obvious example would be that harvest costs are likely to vary with terrain and log size, and labor and equipment costs also may vary. However, it should be noted here that these differences are probably not as large as one might initially believe, since the study deals exclusively with a subset of all harvests, that is, the harvest of even-aged plantation forests. For most of the plantations the log sizes are relatively similar. In addition, the majority of regions that are being examined have somewhat similar, relatively flat terrain. (The major exception is the Pacific Northwest region, which has more mountainous terrain. However, this higher-cost feature is at least partially offset by larger log size and larger harvest volumes, both of which reduce harvest costs. Labor and equipment costs in the Pacific Northwest also tend to be cost offsetting, since possibilities are substantial.) A related consideration is transport distance between harvest site and mill. Again it should be noted that the study addresses plantations that are assumed to be efficiently designed on contiguous lands. It is significant, however, that for a given mill capacity the land area required to service the mill will be inversely related to the biological growth rates experienced within the plantation. Thus regions with the faster-growing species will require a smaller amount of land area in each plantation, and some internal transport cost savings should be realized from this source.

The possibility of systematic regional cost differences may also apply to processing facilities. In principle the technology and capital required for the production of wood pulp or sawnwood or both are available at more or less standard prices worldwide. However, construction costs may vary depending on the accessibility of a particular plantation site or, more relevant to this study, because of the accessibility of the region itself. Thus, for example, regional accessibility considerations, including the availability and skill of the labor force, may result in higher construction and therefore higher mill capital costs in regions such as Amazonia or Borneo. Similar considerations for the labor force may also be relevant for the operation of the mill. On the one hand, the costs of obtaining and keeping technical personnel in isolated locations may be substantial. On the other hand, local labor costs may be low by international standards, and thus some operations, especially in the sawmill, may utilize inexpensive labor. Also, the opportunities for substantial cost-saving, labor-intensive techniques may be present.

A third reason that regional costs may differ from the posited international costs could be found in the presence of trade restrictions that substantially increase local costs. An example would be a local country's restriction on the import of various types of capital equipment required for harvesting, internal transportation, or processing. To the extent that import restrictions raise the local costs of any or all of the various activities necessary to operate an integrated plantation, the costs experienced in the region would be expected to be systematically higher than those incorporated in the model. The restriction could take the form of an import duty, a policy to require the purchase of high-priced, domestically produced equipment, or other similar cost-creating restrictions. It should be noted that costs resulting from artificial restrictions of this type do not reflect the true "economic" or opportunity costs of the productive factors, but instead are "financial" costs induced by institutional factors. Since the focus of the study is the true inherent economic cost, there is no systematic attempt to incorporate restriction-induced financial costs. In this regard two considerations should be noted. First, governmentally imposed restrictions can be imposed or removed at the whim of the government, and thus, while they constitute an appropriate financial consideration for profit-maximizing firms or for individuals making current investment decisions, they are not believed to be an appropriate focus for a long-term analysis of the underlying economics. Second, in the sensitivity analyses the impact of potentially higher costs on the various facets of operation is assessed; the reason for the higher cost is not important for purposes of the analysis. The impact of a 20 percent increase in processing costs will have the same sensitivity effects upon the results whether the higher costs are the result of "true" economic conditions or of "artificial" distortions imposed by governmental policies.

A final reason why costs may be expected to change is the length of time covered by the analysis—as much as eighty years (the rotation periods range from six to eighty years). Also, the analysis looks at the plantation "in perpetuity." Costs of management begin in year one and are largely confined to the early years of the rotation period. However, in an in-perpetuity analysis the management costs will be borne again at the beginning of every rotation. Harvest costs, of course, come at the time of commercial thinnings and final harvest. Thus, for the first rotation the final harvest costs will be occurring only after six to eighty years have passed. Processing costs are incurred at the time of commercial thinning and harvest and, as with harvest cost, will be experienced at very different points through time. Therefore, the same types of uncertainty as to future cost levels are relevant. Finally, international transportation costs will occur only after the production of the final product, again at very different times in the future. Thus, any set of prices or costs used, no matter how precisely they capture the situation of 1979, is bound to become nonrepresentative of actual real prices and costs that will exist in the future. Therefore, the emphasis has been on the reasonableness of current prices and costs as representative of the foreseeable future, rather than on the precision with which they reflect the situation in 1979. Similarly, since actual future prices and costs are unknown, the sensitivity analysis provides a basis for a judgment as to what types of price and cost changes can dramatically affect the economic viability of the plantation, and which are likely to have only small impacts.

Focus of the Analysis

The reactions of PNVs and IRRs to the following changes are examined in the sensitivity analyses: (1) both an increase and a decrease in harvest and internal transport costs; (2) an increase in international transportation costs; (3) an increase in pulp mill processing costs; (4) an increase in sawmill processing costs; and (5) a 2 percent increase in traditional producers' sawtimber stumpage prices and an increase in pulpwood stumpage prices of 1 percent in perpetuity. The first analysis, harvest and internal transport costs, involves both increases and decreases in the cost in order to demonstrate the types of analyses that could be undertaken. For the other analyses, cost increases only are considered.

The typical variation in cost examined is 20 percent. As in the base case discussed in chapter 5, the effect of the changes upon the PNV are assessed using both a 5 and a 10 percent discount rate. The revised IRR is also calculated. The sensitivity analysis could be extended to examine numerous other assumptions, components, and facets of the model, but the cost elements examined here are those that are believed to be most susceptible to change.

Harvest and Internal Transport Costs

This section examines the impacts upon the investment criteria of 20 percent higher and of 20 percent lower harvest and internal transport costs. Tables 6-1 through 6-3 present the estimates for the low-harvest-cost scenario. As expected, all the PNVs and IRRs increase over those of the base case. Using a 5 percent discount rate (table 6-1), the Nordic pulpwood regime PNV improves ($-$$43/hectare versus $-$$100/hectare for the base case shown in table 5-1), but nevertheless remains negative. For the 10 percent discount rate (table 6-2), the same cases and regimes are negative as in the base case (table 5-2), with the exception of the Australian integrated, lower-quality sawtimber and the PNW integrated, high-yield-site regimes. Thus, the negative-PNV regimes at the 10 percent discount rate are three Pacific Northwest cases and the Nordic region (both cases). Tables 6-4 through 6-6 present the findings for the high-harvest-cost scenario. For this scenario, of course, the investment criteria values at the 5 percent discount rate diminish, but not enough to generate a sign different from that of the base case PNVs (table 6-4 as compared with table 5-1). For the 10 percent discount rate, however, four additional plantation regimes' PNVs became negative (table 6-5 as compared with table 5-2). The change to negative PNVs occurred in Australia for two regimes, in New Zealand's discounted price, lower-quality sawtimber integrated regime, and in Gambia-Senegal's eucalyptus pulpwood regime.

Finally, tables 6-7 through 6-9 present estimates of the range and variation of the PNVs and IRRs to a 20 percent increase and decrease in harvest costs. The magnitude of the variation of the PNVs (at 5 percent discount rate) ranges from $+$$57 for the Nordic region to $+$$1,075 for central Brazil, with the larger impacts tending to be on the plantations with large positive PNVs, and particularly on the hardwood plantations. Since the large absolute impacts are on the high-PNV plantations, few PNVs change signs from those of the base case as harvest costs move through the prescribed range.

International Transport Costs

A component of the model that exhibits a great deal of variability across regions in the real world, both at a point in time and also through time, is international transport costs. (See chapter 3 for discussion of factors affecting actual transport costs.)

This study has used a general approach of estimating international transport costs as a function of distance without regard to the particulars that applied to the various regions in 1979. The rationale for this approach is that the transport cost structure is likely to change considerably if large-scale plantations and mills are put into operation. The subsequent volumes and rate structure are likely to change dramatically, approximating to a greater degree the costs experienced on the more heavily traveled routes. This perspective is consistent with the major study objective of estimating the long-term viability of large-scale forest plantation operations.

TABLE 6-1. Low-Harvest-Cost Scenario; Present Net Value; 5 Percent Discount Rate; 1979 Constant Prices in Perpetuity
(1979 U.S. $/hectare)

Region/species	Regime		
	Pulpwood	Integrated, with standard-quality sawtimber	Integrated, with lower-quality sawtimber
North America			
U.S. South			
Pinus taeda, avg.-yield site	2,094	2,821	—
Pinus taeda, high-yield site	3,350	4,246	—
Pacific Northwest			
Pseudotsuga menziesii, avg.-yield site	819	1,068	—
Pseudotsuga menziesii, high-yield site	1,372	2,638	—
South America			
Brazil, Amazonia			
Pinus cribaea	3,666	4,799	4,096
Gmelina spp.	3,283	3,992	—
Brazil, Central			
Eucalyptus spp.	4,502	5,102	—
Brazil, Southern			
Pinus taeda	4,316	5,642	4,880
Chile			
Pinus radiata	4,418	5,164	3,860
Oceania			
Australia			
Pinus radiata	2,688	2,725	1,962
New Zealand			
Pinus radiata	3,671	5,045	3,494
Africa			
South Africa			
Pinus patula	3,588	4,270	3,325
Gambia-Senegal			
Gmelina spp.	2,861	3,298	—
Eucalyptus spp.	2,530	4,002	—
Europe			
Nordic			
Picea abies	−43	250	—
Asia			
Borneo			
Pinus caribaea	2,318	2,937	2,418

Two alternative international transport cost scenarios are examined. The first scenario assumes that international transport costs increase 20 percent *for all regions*. In such a situation, part of the higher transport costs would be borne by the consumer in the form of higher prices in world markets, while the remainder of the increased transport cost would be borne by stumpage prices everywhere. Since the absolute amount of the higher cost will depend upon initial transport costs, the higher transport costs will accentuate the effects of location, and each region's costs would not increase equally. Advantageously located low-transport-cost regions would experience the smallest absolute increase in total costs, and thus their competitive position vis-à-vis less advantageously situated competitors would be enhanced. For example, the effect of a 20 percent transport cost increase on PNVs of the U.S. South or Nordic region, which are near their final markets, would be quite small. By contrast, the effect on PNVs of Chile or New Zealand, for example, would be relatively large.

Furthermore, as explained below, the higher transport costs are translated, in part, into higher prices in final markets, so that the higher international transport costs for some regional producers could be more than offset by the higher product prices in the

TABLE 6-2. Low-Harvest-Cost Scenario; Present Net Value; 10 Percent Discount Rate; 1979 Constant Prices in Perpetuity
(1979 U.S. $/hectare)

	Regime		
Region/species	Pulpwood	Integrated, with standard-quality sawtimber	Integrated, with lower-quality sawtimber
North America			
U.S. South			
Pinus taeda, avg.-yield site	282	363	—
Pinus taeda, high-yield site	587	682	—
Pacific Northwest			
Pseudotsuga menziesii, avg.-yield site	−251	−328	—
Pseudotsuga menziesii, high-yield site	−59	9	—
South America			
Brazil, Amazonia			
Pinus cribaea	1,065	1,516	1,275
Gmelina spp.	1,270	1,389	—
Brazil, Central			
Eucalyptus spp.	1,596	1,351	—
Brazil, Southern			
Pinus taeda	1,135	1,631	1,403
Chile			
Pinus radiata	1,289	1,076	772
Oceania			
Australia			
Pinus radiata	330	185	21
New Zealand			
Pinus radiata	581	948	497
Africa			
South Africa			
Pinus patula	1,030	1,015	763
Gambia-Senegal			
Gmelina spp.	908	966	—
Eucalyptus spp.	581	1,090	—
Europe			
Nordic			
Picea abies	−492	−397	—
Asia			
Borneo			
Pinus caribaea	461	716	559

TABLE 6-3. Low-Harvest-Cost Scenario, Internal Rate of Return, 1979 Constant Prices in Perpetuity
(percent)

Region/species	Pulpwood	Regime Integrated, with standard-quality sawtimber	Integrated, with lower-quality sawtimber
North America			
U.S. South			
Pinus taeda, avg.-yield site	12.79	13.07	—
Pinus taeda, high-yield site	14.74	14.81	—
Pacific Northwest			
Pseudotsuga menziesii, avg.-yield site	7.57	7.31	—
Pseudotsuga menziesii, high-yield site	9.47	10.06	—
South America			
Brazil, Amazonia			
Pinus cribaea	19.25	22.14	21.23
Gmelina spp.	31.39	26.78	—
Brazil, Central			
Eucalyptus spp.	23.04	17.68	—
Brazil, Southern			
Pinus taeda	16.85	19.26	18.66
Chile			
Pinus radiata	25.84	18.29	17.00
Oceania			
Australia			
Pinus radiata	12.15	11.04	10.14
New Zealand			
Pinus radiata	12.95	14.21	12.55
Africa			
South Africa			
Pinus patula	20.50	18.86	17.63
Gambia-Senegal			
Gmelina spp.	20.31	20.04	—
Eucalyptus spp.	14.20	16.33	—
Europe			
Nordic			
Picea abies	4.84	5.90	—
Asia			
Borneo			
Pinus caribaea	14.16	16.35	15.44

final markets. As will be seen, the higher international transport costs actually improve the PNV of the Nordic pulpwood plantations, since the higher wood pulp prices more than offset the Nordic region's relatively small cost increase resulting from the higher international transport costs.

A higher level of international transport costs will affect both the delivered price and shadow stumpage price. This scenario is examined in tables 6-10 to 6-12. In the absence of detailed knowledge of the relevant demand and supply elasticities, it is assumed that one-half of the effect of higher transport costs would fall on the European market, thereby raising delivered prices, and one-half of the effect would fall on the Pacific Northwest, thereby lowering stumpage prices (see appendix D, tables D-4 and D-5, for high-transportation-cost stumpage prices). Given this adjustment, the delivered prices in the other major world markets were determined, and new regional stumpage prices were calculated to recognize the high transport costs and new structure of international prices.

For example, the total increase in transport costs for wood pulp between the Pacific Northwest (PNW) and Europe was $10.2 per metric ton. This was translated into an increase of the European market price from $450 to $455.1 per metric ton, with the re-

TABLE 6-4. High-Harvest-Cost Scenario; Present Net Value; 5 Percent Discount Rate; 1979 Constant Prices in Perpetuity

(1979 U.S. $/hectare)

Region/species	Regime		
	Pulpwood	Integrated, with standard-quality sawtimber	Integrated, with lower-quality sawtimber
North America			
U.S. South			
Pinus taeda, avg.-yield site	1,402	2,128	—
Pinus taeda, high-yield site	2,309	3,239	—
Pacific Northwest			
Pseudotsuga menziesii, avg.-yield site	413	737	—
Pseudotsuga menziesii, high-yield site	783	1,858	—
South America			
Brazil, Amazonia			
Pinus caribaea	2,508	3,360	2,657
Gmelina spp.	1,777	2,376	—
Brazil, Central			
Eucalyptus spp.	2,411	2,952	—
Brazil, Southern			
Pinus taeda	2,868	3,788	3,026
Chile			
Pinus radiata	2,880	3,855	2,553
Oceania			
Australia			
Pinus radiata	1,321	1,557	794
New Zealand			
Pinus radiata	2,136	3,192	1,641
Africa			
South Africa			
Pinus patula	2,514	3,185	2,241
Gambia-Senegal			
Gmelina spp.	1,717	1,846	—
Eucalyptus spp.	1,127	2,522	—
Europe			
Nordic			
Picea abies	−158	58	—
Asia			
Borneo			
Pinus caribaea	1,384	1,792	1,274

TABLE 6-5. High-Harvest-Cost Scenario; Present Net Value; 10 Percent Discount Rate; 1979 Constant Prices in Perpetuity
(1979 U.S. $/hectare)

		Regime	
Region/species	Pulpwood	Integrated, with standard-quality sawtimber	Integrated, with lower-quality sawtimber
North America			
U.S. South			
Pinus taeda, avg.-yield site	99	189	—
Pinus taeda, high-yield site	316	428	—
Pacific Northwest			
Pseudotsuga menziesii, avg.-yield site	−306	−357	—
Pseudotsuga menziesii, high-yield site	−178	−119	—
South America			
Brazil, Amazonia			
Pinus caribaea	634	948	708
Gmelina spp.	618	714	—
Brazil, Central			
Eucalyptus spp.	690	504	—
Brazil, Southern			
Pinus taeda	596	894	666
Chile			
Pinus radiata	750	740	437
Oceania			
Australia			
Pinus radiata	−133	−165	−328
New Zealand			
Pinus radiata	107	355	−95
Africa			
South Africa			
Pinus patula	665	673	421
Gambia-Senegal			
Gmelina spp.	456	421	—
Eucalyptus spp.	−21	461	—
Europe			
Nordic			
Picea abies	−502	−438	—
Asia			
Borneo			
Pinus caribaea	143	288	131

maining $5.1 being appropriately converted and then reduced from the PNW stumpage prices. The market prices in Japan and on the East Coast of the United States were adjusted to make them consistent with both the increased European price and the high international transport costs. The increased prices for wood pulp and lumber in the three major world markets are shown below.

World Market Prices
(1979 U.S. dollars per metric ton)

	Base case		
	Japan	U.S. East	Europe
Wood pulp	431.6	438.4	450.0
Lumber	295.5	306.4	325.0
	International transport costs increased 20 percent		
Wood pulp	433.1	441.3	455.1
Lumber	297.9	311.0	333.2

The second international transport cost scenario assumes that the international transport cost of each individual region increases by 20 percent, while the rest of the world's plantation international-transport costs remain unchanged. In this situation the prices of products in final markets would not be affected. Thus the entire impact of the higher transport cost for the region would be borne by the stumpage prices in that region. In effect, the higher transport costs required to get the region's product to world markets results in a deterioration in the region's competitive position, which is reflected in depressed stumpage prices.

Tables 6-13 through 6-15 examine the sensitivity of the PNV and IRR to a 20 percent increase in international transport costs for

TABLE 6-6. High-Harvest-Cost Scenario; Internal Rate of Return; 1979 Constant Prices in Perpetuity
(percent)

	Regime		
Region/species	Pulpwood	Integrated, with standard-quality sawtimber	Integrated, with lower-quality sawtimber
North America			
U.S. South			
Pinus taeda, avg.-yield site	11.13	11.76	—
Pinus taeda, high-yield site	12.97	13.35	—
Pacific Northwest			
Pseudotsuga menziesii, avg.-yield site	6.57	6.80	—
Pseudotsuga menziesii, high-yield site	8.11	9.13	—
South America			
Brazil, Amazonia			
Pinus caribaea	16.34	18.55	17.11
Gmelina spp.	22.91	19.86	—
Brazil, Central			
Eucalyptus spp.	16.78	13.18	—
Brazil, Southern			
Pinus taeda	14.12	15.63	14.67
Chile			
Pinus radiata	20.56	16.62	14.83
Oceania			
Australia			
Pinus radiata	9.03	8.95	7.52
New Zealand			
Pinus radiata	10.65	11.83	9.40
Africa			
South Africa			
Pinus patula	18.01	16.44	14.76
Gambia-Senegal			
Gmelina spp.	16.21	14.79	—
Eucalyptus spp.	9.82	12.94	—
Europe			
Nordic			
Picea abies	4.35	5.22	—
Asia			
Borneo			
Pinus caribaea	11.52	12.90	11.49

TABLE 6-7. Sensitivity of Present Net Value to Harvest Cost Variations, 5 percent Discount Rate
(1979 U.S. $/hectare)

Region/species	Regime		
	Pulpwood	Integrated, with standard-quality sawtimber	Integrated, with lower-quality sawtimber
North America			
U.S. South			
Pinus taeda, avg.-yield site	±346	±347	—
Pinus taeda, high-yield site	±520	+504	—
Pacific Northwest			
Pseudotsuga menziesii, avg.-yield site	±203	±165	—
Pseudotsuga menziesii, high-yield site	+295	±390	—
South America			
Brazil, Amazonia			
Pinus cribaea	±579	±719	±720
Gmelina spp.	±753	±808	—
Brazil, Central			
Eucalyptus spp.	±1,046	±1,075	—
Brazil, Southern			
Pinus taeda	±724	±927	±927
Chile			
Pinus radiata	±769	±655	±654
Oceania			
Australia			
Pinus radiata	±683	±584	±584
New Zealand			
Pinus radiata	±768	±927	±927
Africa			
South Africa			
Pinus patula	±537	±543	±542
Gambia-Senegal			
Gmelina spp.	±572	±676	—
Eucalyptus spp.	±702	±740	—
Europe			
Nordic			
Picea abies	±57	±96	—
Asia			
Borneo			
Pinus caribaea	±467	±573	±572

each region and regime in isolation, thus assuming that all the other regions and regimes are experiencing base case international transport costs.

Of the thirty-two regimes using a 5 percent discount rate (table 6-13), the signs of all the PNVs are unchanged from those of the base case (table 5-1). Only the same regional pulpwood regime (Nordic region) has a negative PNV, as in the base case. As would be expected, all the PNVs were reduced, with the largest reductions experienced by the plantations that had the largest absolute cost increases, for example, those farthest from world markets. For the 10 percent discount rate (table 6-14), the higher transport costs result in some PNVs that are positive in the base case (table 5-2) becoming negative, namely, two additional Australian regimes and New Zealand's integrated (with lower-quality sawtimber) regime.

Pulp Mill Processing Costs

Of all the elements investigated by sensitivity analysis, increases in the costs of pulp mill processing, that is, in production of bleached kraft wood pulp, have the largest impact on the representative plantations' estimated PNVs and IRRs. In principle there are prima facie reasons supporting the theory that the economic costs of pulp mill processing ought to be similar throughout the world. The technology is relatively standard and inflexible. All regions, in principle, have

TABLE 6-8. Sensitivity of Present Net Value to Harvest Cost Variations, 10 Percent Discount Rate
(1979 U.S. $/hectare)

Region/species	Pulpwood	Regime Integrated, with standard-quality sawtimber	Integrated, with lower-quality sawtimber
North America			
U.S. South			
Pinus taeda, avg.-yield site	±92	±87	—
Pinus taeda, high-yield site	±135	±127	—
Pacific Northwest			
Pseudotsuga menziesii, avg.-yield site	±27	±14	—
Pseudotsuga menziesii, high-yield site	±59	±64	—
South America			
Brazil, Amazonia			
Pinus caribaea	±215	±284	±284
Gmelina spp.	±326	±337	—
Brazil, Central			
Eucalyptus spp.	±453	±423	—
Brazil, Southern			
Pinus taeda	±270	±369	±368
Chile			
Pinus radiata	±270	±168	±167
Oceania			
Australia			
Pinus radiata	±231	±175	±174
New Zealand			
Pinus radiata	±237	±296	±296
Africa			
South Africa			
Pinus patula	±183	±171	±171
Gambia-Senegal			
Gmelina spp.	±226	±273	—
Eucalyptus spp.	±301	±315	—
Europe			
Nordic			
Picea abies	±5	±20	—
Asia			
Borneo			
Pinus caribaea	±159	±214	±214

TABLE 6-9. Sensitivity of Present Net Value to Harvest Cost Variations, Internal Rate of Return
(percent)

		Regime	
Region/species	Pulpwood	Integrated, with standard-quality sawtimber	Integrated, with lower-quality sawtimber
North America			
U.S. South			
Pinus taeda, avg.-yield site	+.77/−.89	+.62/−.69	—
Pinus taeda, high-yield site	+.83/−.94	+1.69/−.77	—
Pacific Northwest			
Pseudotsuga menziesii, avg.-yield site	+1.46/−.54	+.24/−.27	—
Pseudotsuga menziesii, high-yield site	+.62/−.74	+.44/−.49	—
South America			
Brazil, Amazonia			
Pinus caribaea	+1.36/−1.55	+1.70/−1.89	+1.88/−2.14
Gmelina spp.	+3.86/−4.62	+3.24/−3.68	—
Brazil, Central			
Eucalyptus spp.	+2.88/−3.38	+2.14/−2.36	—
Brazil, Southern			
Pinus taeda	+1.28/−1.45	+1.73/−1.90	+1.87/−2.12
Chile			
Pinus radiata	+2.45/−2.83	+.79/−.88	+.99/−1.18
Oceania			
Australia			
Pinus radiata	+1.47/−1.65	+.98/−1.11	+1.20/−1.42
New Zealand			
Pinus radiata	+1.05/−1.25	+1.10/−1.28	+1.42/−1.73
Africa			
South Africa			
Pinus patula	+1.16/−1.33	+1.17/−1.25	+1.36/−1.51
Gambia-Senegal			
Gmelina spp.	+1.89/−2.21	+2.52/−2.73	—
Eucalyptus spp.	+2.04/−2.34	+1.62/−1.77	—
Europe			
Nordic			
Picea abies	+.23/−.26	+.33/−.35	—
Asia			
Borneo			
Pinus caribaea	+1.22/−1.42	+1.62/−1.83	+1.83/−2.12

access to both the technology and capital that are available in international markets. Labor costs are a relatively minor portion of the production process, and therefore regional differences in labor wages are not likely to have large effects on total costs. Finally, much of the labor input consists of trained technical people, and the regional wage differences here are smaller than for unskilled labor. Major differences in economic costs by region might originate with different mill construction costs reflecting local conditions. Also, local regulations and restrictions could affect the financial costs of the construction or the operation of a wood pulp mill. (These, of course, would not reflect true economic costs.) The basic processing costs built into the model are those applicable to the United States and the Nordic countries. These are probably close to the current low bound of costs likely to be experienced in any region internationally because of the accessibility and developed infrastructures of these regions. Thus, for the nontraditional producers, the U.S./Nordic pulp processing costs would probably be the norm to strive toward. Actual deviations from these costs appear likely to be on the high side.

The sensitivity of the PNVs, especially the pulpwood PNVs, to increasing pulp processing costs 20 percent is dramatic, as seen

TABLE 6-10. High-Transportation-Cost Scenario; World Market Prices Adjusted; Present Net Value; 5 Percent Discount Rate; 1979 Constant Prices in Perpetuity
(1979 U.S. $/hectare)

	Regime		
Region/species	Pulpwood	Integrated, with standard-quality sawtimber	Integrated, with lower-quality sawtimber
North America			
U.S. South			
Pinus taeda, avg.-yield site	1,727	2,397	—
Pinus taeda, high-yield site	2,799	3,631	—
Pacific Northwest			
Pseudotsuga menziesii, avg.-yield site	523	692	—
Pseudotsuga menziesii, high-yield site	942	1,849	—
South America			
Brazil, Amazonia			
Pinus caribaea	3,010	3,902	3,184
Gmelina spp.	2,291	2,862	—
Brazil, Central			
Eucalyptus spp.	3,125	3,578	—
Brazil, Southern			
Pinus taeda	3,495	4,509	3,732
Chile			
Pinus radiata	3,181	3,923	2,575
Oceania			
Australia			
Pinus radiata	1,701	1,728	874
New Zealand			
Pinus radiata	2,516	3,381	1,647
Africa			
South Africa			
Pinus patula	2,944	3,443	2,433
Gambia-Senegal			
Gmelina spp.	2,229	2,483	—
Eucalyptus spp.	1,755	3,051	—
Europe			
Nordic			
Picea abies	−95	156	—
Asia			
Borneo			
Pinus caribaea	1,616	2,055	1,506

TABLE 6-11. High-Transportation-Cost Scenario; World Market Prices Adjusted; Present Net Value; 10 Percent Discount Rate; 1979 Constant Prices in Perpetuity
(1979 U.S. $/hectare)

Region/species	Pulpwood	Regime Integrated, with standard-quality sawtimber	Integrated, with lower-quality sawtimber
North America			
U.S. South			
Pinus taeda, avg.-yield site	185	262	—
Pinus taeda, high-yield site	444	534	—
Pacific Northwest			
Pseudotsuga menziesii, avg.-yield site	−291	−361	—
Pseudotsuga menziesii, high-yield site	−146	−115	—
South America			
Brazil, Amazonia			
Pinus caribaea	821	1,167	922
Gmelina spp.	841	923	—
Brazil, Central			
Eucalyptus spp.	1,000	765	—
Brazil, Southern			
Pinus taeda	829	1,192	960
Chile			
Pinus radiata	864	766	452
Oceania			
Australia			
Pinus radiata	2	−96	−279
New Zealand			
Pinus radiata	225	430	−74
Africa			
South Africa			
Pinus patula	811	765	495
Gambia-Senegal			
Gmelina spp.	659	645	—
Eucalyptus spp.	248	690	—
Europe			
Nordic			
Picea abies	−496	−416	—
Asia			
Borneo			
Pinus caribaea	222	393	227

in tables 6-16 through 6-18. Using a 5 percent discount rate (table 6-16), ten of the sixteen pulpwood regimes generate negative PNVs, including, importantly, some of the base case's best-performing plantations of South America, Oceania, and Africa. For all the integrated regimes the signs remain unchanged (two of the sixteen cases have negative PNVs).

The effects are even more dramatic when a 10 percent discount rate is used (table 6-17). In this case all of the sixteen pulpwood plantations have negative PNVs. Ten of the sixteen integrated regimes with standard-quality sawtimber have negative PNVs using a 10 percent rate. The IRRs are similarly depressed. The integrated plantations are impacted in two ways. First, their pulpwood stumpage prices are, of course, reduced—reflecting the higher pulp processing costs. In addition, the sawtimber stumpage prices are reduced, reflecting the decline in value of that portion of the sawtimber stumpage, for example, residuals, that is eventually used as pulpwood. Through this mechanism, sawtimber stumpage prices as well as pulpwood prices are depressed.

The effect on plantation economics of increasing this cost element is so great for several reasons. First, pulp mill processing costs constitute a large proportion—well over 50 percent—of final product costs. Therefore, a 20 percent increase in pulp milling costs

TABLE 6-12. High-Transportation-Cost Scenario; World Market Prices Adjusted; Internal Rate of Return; 1979 Constant Prices in Perpetuity
(percent)

Region/species	Regime		
	Pulpwood	Integrated, with standard-quality sawtimber	Integrated, with lower-quality sawtimber
North America			
U.S. South			
Pinus taeda, avg.-yield site	11.97	12.35	—
Pinus taeda, high-yield site	13.87	14.02	—
Pacific Northwest			
Pseudotsuga menziesii, avg.-yield site	6.88	6.72	—
Pseudotsuga menziesii, high-yield site	8.53	9.15	—
South America			
Brazil, Amazonia			
Pinus caribaea	17.69	20.09	18.83
Gmelina spp.	26.17	22.33	—
Brazil, Central			
Eucalyptus spp.	19.15	14.73	—
Brazil, Southern			
Pinus taeda	15.39	17.24	16.45
Chile			
Pinus radiata	21.94	16.81	15.00
Oceania			
Australia			
Pinus radiata	10.01	9.39	7.86
New Zealand			
Pinus radiata	11.30	12.19	9.53
Africa			
South Africa			
Pinus patula	19.09	17.23	15.57
Gambia-Senegal			
Gmelina spp.	18.21	17.15	—
Eucalyptus spp.	11.93	14.27	—
Europe			
Nordic			
Picea abies	4.64	5.58	—
Asia			
Borneo			
Pinus caribaea	12.25	13.87	12.52

TABLE 6-13. High-Transportation-Cost Scenario; World Market Prices Unchanged; Present Net Value; 5 Percent Discount Rate; 1979 Constant Prices in Perpetuity
(1979 U.S. $/hectare)

		Regime	
Region/species	Pulpwood	Integrated, with standard-quality sawtimber	Integrated, with lower-quality sawtimber
North America			
U.S. South			
Pinus taeda, avg.-yield site	1,649	2,329	—
Pinus taeda, high-yield site	2,683	3,532	—
Pacific Northwest			
Pseudotsuga menziesii, avg.-yield site	430	583	—
Pseudotsuga menziesii, high-yield site	807	1,628	—
South America			
Brazil, Amazonia			
Pinus caribaea	2,749	3,585	2,837
Gmelina spp.	2,025	2,552	—
Brazil, Central			
Eucalyptus spp.	2,755	3,158	—
Brazil, Southern			
Pinus taeda	3,170	4,121	3,311
Chile			
Pinus radiata	3,090	3,681	1,926
Oceania			
Australia			
Pinus radiata	1,620	1,636	770
New Zealand			
Pinus radiata	2,414	3,224	1,463
Africa			
South Africa			
Pinus patula	2,662	3,146	2,056
Gambia-Senegal			
Gmelina spp.	1,934	2,159	—
Eucalyptus spp.	1,393	2,637	—
Europe			
Nordic			
Picea abies	−120	118	—
Asia			
Borneo			
Pinus caribaea	1,553	1,980	1,429

translates into a relatively large percentage increase of final costs. The effect of a large increase in costs is a large absolute reduction in plantation stumpage prices, which thereby reduces the PNVs. The effect is greatest on the plantations of the tropics and the Southern Hemisphere, because the high PNVs of these plantations tend to be the result of very rapid biological growth, which offsets the effects of stumpage prices that are generally lower than those of the temperate Northern Hemisphere. Since the stumpage prices are lower initially, a decrease of a set absolute amount translates into a large percentage reduction. This situation is seen in its extreme for the tropical hardwood plantations, which have low initial stumpage prices. Their PNVs are the most adversely affected in this sensitivity scenario.

In summary, the sensitivity analysis has shown that the impact of a 20 percent increase in wood pulp processing costs on PNVs is large and important. The PNVs for pulpwood plantations decline precipitously, and the PNVs of integrated regimes are also considerably reduced. This finding is in marked contrast to the sensitivity analysis of the other cost components of the production process, which produced only a modest impact upon the PNVs.

This result is significant not only because it demonstrates the importance of holding

TABLE 6-14. High-Transportation-Cost Scenario; World Market Prices Unchanged; Present Net Value; 10 Percent Discount Rate; 1979 Constant Prices in Perpetuity
(1979 U.S. $/hectare)

Region/species	Regime		
	Pulpwood	Integrated, with standard-quality sawtimber	Integrated, with lower-quality sawtimber
North America			
U.S. South			
Pinus taeda, avg.-yield site	165	245	—
Pinus taeda, high-yield site	415	510	—
Pacific Northwest			
Pseudotsuga menziesii, avg.-yield site	−304	−371	—
Pseudotsuga menziesii, high-yield site	−173	−149	—
South America			
Brazil, Amazonia			
Pinus caribaea	724	1,047	791
Gmelina spp.	725	798	—
Brazil, Central			
Eucalyptus spp.	839	610	—
Brazil, Southern			
Pinus taeda	708	1,045	803
Chile			
Pinus radiata	833	708	300
Oceania			
Australia			
Pinus radiata	−24	−121	−306
New Zealand			
Pinus radiata	193	382	−130
Africa			
South Africa			
Pinus patula	715	676	385
Gambia-Senegal			
Gmelina spp.	542	521	—
Eucalyptus spp.	93	518	—
Europe			
Nordic			
Picea abies	−499	−424	—
Asia			
Borneo			
Pinus caribaea	201	367	199

TABLE 6-15. High-Transportation-Cost-Scenario; World Market Prices Unchanged; Internal Rate of Return; 1979 Constant Prices in Perpetuity
(percent)

	Regime		
Region/species	Pulpwood	Integrated, with standard-quality sawtimber	Integrated, with lower-quality sawtimber
North America			
U.S. South			
Pinus taeda, avg.-yield site	11.79	12.22	—
Pinus taeda, high-yield site	13.68	13.88	—
Pacific Northwest			
Pseudotsuga menziesii, avg.-yield site	6.62	6.52	—
Pseudotsuga menziesii, high-yield site	8.18	8.86	—
South America			
Brazil, Amazonia			
Pinus caribaea	17.01	19.34	17.90
Gmelina spp.	24.54	21.05	—
Brazil, Central			
Eucalyptus spp.	17.96	13.89	—
Brazil, Southern			
Pinus taeda	14.75	16.53	15.61
Chile			
Pinus radiata	21.64	16.52	13.81
Oceania			
Australia			
Pinus radiata	9.83	9.23	7.60
New Zealand			
Pinus radiata	11.13	11.98	9.15
Africa			
South Africa			
Pinus patula	18.39	16.62	14.62
Gambia-Senegal			
Gmelina spp.	17.09	15.97	—
Eucalyptus spp.	10.75	13.31	—
Europe			
Nordic			
Picea abies	4.52	5.44	—
Asia			
Borneo			
Pinus caribaea	12.06	13.65	12.25

processing costs down, but also because the wood pulp regime has been viewed as providing the floor with respect to plantation returns. If lumber markets are soft or if the plantation's lumber quality is unacceptable, the plantation must revert to the pulpwood operation. However, as the above sensitivity analysis shows, the economic viability of a pulpwood plantation is critically dependent upon the costs of wood pulp processing in the associated mill. If the regional pulp mill costs are only 20 percent higher than those experienced in other regions, the advantages of rapid biological growth may be partially or wholly negated through a precipitous decline in the stumpage price.

Sawmill Processing Costs

The findings of the sensitivity analysis for sawmill processing costs are found in tables 6-19 through 6-21. It will be noted that the 20 percent increase in sawtimber processing costs does not affect and therefore has no impact on the PNV base case estimates for the pulpwood regimes (table 5-1). Also, the effect of higher sawmill processing costs upon the estimated PNVs of all integrated regimes is dampened, since the pulpwood component of the integrated operation is unaffected and therefore dilutes the sawmill processing impact. Probably the most important effect is that with the higher sawmill

TABLE 6-16. High-Pulp-Processing-Cost Scenario; Present Net Value; 5 Percent Discount Rate; 1979 Constant Prices in Perpetuity
(1979 U.S. $/hectare)

	Regime		
Region/species	Pulpwood	Integrated, with standard-quality sawtimber	Integrated, with lower-quality sawtimber
North America			
U.S. South			
Pinus taeda, avg.-yield site	253	1,467	—
Pinus taeda, high-yield site	597	2,298	—
Pacific Northwest			
Pseudotsuga menziesii, avg.-yield site	−373	618	—
Pseudotsuga menziesii, high-yield site	−361	1,418	—
South America			
Brazil, Amazonia			
Pinus caribaea	262	1,838	1,134
Gmelina spp.	−1,680	291	—
Brazil, Central			
Eucalyptus spp.	−2,391	729	—
Brazil, Southern			
Pinus taeda	61	1,799	1,037
Chile			
Pinus radiata	310	3,086	1,784
Oceania			
Australia			
Pinus radiata	−935	630	−130
New Zealand			
Pinus radiata	−840	2,050	506
Africa			
South Africa			
Pinus patula	432	2,371	1,426
Gambia-Senegal			
Gmelina spp.	−909	259	—
Eucalyptus spp.	−2,092	763	—
Europe			
Nordic			
Picea abies	−380	−134	—
Asia			
Borneo			
Pinus caribaea	−426	594	76

TABLE 6-17. High-Pulp-Processing-Cost Scenario; Present Net Value; 10 Percent Discount Rate; 1979 Constant Prices in Perpetuity
(1979 U.S. $/hectare)

	Regime		
Region/species	Pulpwood	Integrated, with standard-quality sawtimber	Integrated, with lower-quality sawtimber
North America			
U.S. South			
Pinus taeda, avg.-yield site	−190	2	—
Pinus taeda, high-yield site	−111	161	—
Pacific Northwest			
Pseudotsuga menziesii, avg.-yield site	−413	−368	—
Pseudotsuga menziesii, high-yield site	−409	−199	—
South America			
Brazil, Amazonia			
Pinus caribaea	−201	331	91
Gmelina spp.	−879	−180	—
Brazil, Central			
Eucalyptus spp.	−1,388	−438	—
Brazil, Southern			
Pinus taeda	−449	59	−169
Chile			
Pinus radiata	−93	536	232
Oceania			
Australia			
Pinus radiata	−840	−486	−648
New Zealand			
Pinus radiata	−811	−24	−473
Africa			
South Africa			
Pinus patula	−42	397	144
Gambia-Senegal			
Gmelina spp.	−580	−296	—
Eucalyptus spp.	−1,403	−306	—
Europe			
Nordic			
Picea abies	−522	−488	—
Asia			
Borneo			
Pinus caribaea	−471	−179	−336

costs, only four of the sixteen regimes (namely, the U.S. South average-yield site, Pacific Northwest high-yield site, southern Brazil, and the Nordic region) would operate an integrated regime as compared with all for the base case. A modest increase in non-wood-sawmilling costs, therefore, results in depressing the returns to sawtimber and a shift to the now-higher-return pulpwood plantation regime.

Increase in all Stumpage Prices

All of the previous scenarios have assumed constant final-product and implicit stumpage prices indefinitely into the future. However, as noted earlier, stumpage prices, at least for sawtimber stumpage, have been experiencing a long-term upward trend for at least a century. Also, the earlier analysis presented reasons for expecting that this trend

TABLE 6-18. **High-Pulp-Processing-Cost Scenario; Internal Rate of Return; 1979 Constant Prices in Perpetuity**
(percent)

Region/species	Regime		
	Pulpwood	Integrated, with standard-quality sawtimber	Integrated, with lower-quality sawtimber
North America			
U.S. South			
Pinus taeda, avg.-yield site	6.78	10.02	—
Pinus taeda, high-yield site	8.40	11.38	—
Pacific Northwest			
Pseudotsuga menziesii, avg.-yield site	1.92	6.59	—
Pseudotsuga menziesii, high-yield site	1.78	8.43	—
South America			
Brazil, Amazonia			
Pinus caribaea	6.89	13.39	11.07
Gmelina spp.	—	7.11	—
Brazil, Central			
Eucalyptus spp.	—	7.11	—
Brazil, Southern			
Pinus taeda	5.30	10.34	8.67
Chile			
Pinus radiata	7.88	15.28	12.95
Oceania			
Australia			
Pinus radiata	—	6.70	4.54
New Zealand			
Pinus radiata	—	9.86	6.58
Africa			
South Africa			
Pinus patula	9.12	13.97	11.76
Gambia-Senegal			
Gmelina spp.	—	6.40	—
Eucalyptus spp.	—	7.78	—
Europe			
Nordic			
Picea abies	2.89	4.48	—
Asia			
Borneo			
Pinus caribaea	1.25	7.96	5.46

may continue until the transition to a managed world forest is completed.

Two price rise scenarios are examined. For both it is assumed that stumpage price would rise in perpetuity. Although this assumption is unrealistic, it does not seriously distort the final PNV estimates.[1]

1 Percent Growth in All Stumpage Prices

Of course, the increase of the real price of stumpage will, *ceteris paribus*, increase the estimates of economic returns. The results examined and compared with the assumption of price increases in perpetuity. It was found that because of the effect of discounting, the assumption of an indefinitely rising price generates only a small upward bias in the PNVs compared with an assumption of a leveling off after fifty years.

[1] What would happen under the assumption that the price rise continues for a period of time, after which constant prices obtain? This type of price behavior was

TABLE 6-19. High-Lumber-Production-Cost Scenario; Present Net Value, 5 Percent Discount Rate; 1979 Constant Prices in Perpetuity (Base Stumpage Prices for Pulpwood)
(1979 U.S. $/hectare)

	Regime		
Region/species	Pulpwood	Integrated, with standard-quality sawtimber	Integrated, with lower-quality sawtimber
North America			
U.S. South			
Pinus taeda, avg.-yield site	1,748	1,822	—
Pinus taeda, high-yield site	2,830	2,797	—
Pacific Northwest			
Pseudotsuga menziesii, avg.-yield site	616	271	—
Pseudotsuga menziesii, high-yield site	1,077	1,147	—
South America			
Brazil, Amazonia			
Pinus caribaea	3,087	3,054	2,351
Gmelina spp.	2,530	1,700	—
Brazil, Central			
Eucalyptus spp.	3,456	1,566	—
Brazil, Southern			
Pinus taeda	3,592	3,605	2,843
Chile			
Pinus radiata	3,649	2,612	1,309
Oceania			
Australia			
Pinus radiata	2,005	919	157
New Zealand			
Pinus radiata	2,903	1,636	87
Africa			
South Africa			
Pinus patula	3,051	2,351	1,406
Gambia-Senegal			
Gmelina spp.	2,289	1,585	—
Eucalyptus spp.	1,828	1,465	—
Europe			
Nordic			
Picea abies	−100	40	—
Asia			
Borneo			
Pinus caribaea	1,851	1,534	1,016

of a 1 percent increase in perpetuity are found in tables 6-22 and 6-23. Under this scenario all PNVs are positive using a 5 percent discount rate.

2 Percent Growth in Sawtimber Prices and 1 Percent Growth for Pulpwood

The last price scenario is designed to examine the economics of plantation forestry in a world where stumpage prices are rising, but where certain "high-quality" sawtimber stumpage prices are rising even faster.

Tables 6-24 through 6-27 present the results of this scenario. The effect of higher stumpage is, of course, to increase the returns and hence the values of the PNV and IRR.

Selective use of the study's results can provide insights into some outstanding issues in forest management. For example, Fenton (1972) has argued for extensive pruning of New Zealand's young trees on the basis that the future market for high-quality stumpage will justify the increase in current expenditures. A comparison of tables 6-22 and 6-25 reveals that if the annual growth rate of the price of sawtimber stumpage is raised by 1 percent (from 1 percent to 2 percent), the PNV of New Zealand stumpage would be increased by $3,235 (from $6,292 per hectare to $9,527). If it is believed that pruning would sufficiently increase the quality of the stumpage to result in a 2 percent long-term growth in price as opposed to only 1 percent in the absence of pruning, the additional value would almost surely justify the additional cost.

More generally, it is sometimes maintained that temperate-forest sawtimber is of inherently higher quality and is therefore

TABLE 6-20. High-Lumber-Production-Cost Scenario; Present Net Value, 10 Percent Discount Rate; 1979 Constant Prices in Perpetuity (Base Stumpage Prices for Pulpwood)
(1979 U.S. $/hectare)

Region/species	Regime		
	Pulpwood	Integrated, with standard-quality sawtimber	Integrated, with lower-quality sawtimber
North America			
U.S. South			
Pinus taeda, avg.-yield site	190	166	—
Pinus taeda, high-yield site	452	392	—
Pacific Northwest			
Pseudotsuga menziesii, avg.-yield site	−278	−399	—
Pseudotsuga menziesii, high-yield site	−119	−215	—
South America			
Brazil, Amazonia			
Pinus caribaea	850	881	641
Gmelina spp.	944	499	—
Brazil, Central			
Eucalyptus spp.	1,143	128	—
Brazil, Southern			
Pinus taeda	865	931	703
Chile			
Pinus radiata	1,019	466	163
Oceania			
Australia			
Pinus radiata	99	−252	−415
New Zealand			
Pinus radiata	344	−70	−520
Africa			
South Africa			
Pinus patula	847	476	224
Gambia-Senegal			
Gmelina spp.	682	371	—
Eucalyptus spp.	280	66	—
Europe			
Nordic			
Picea abies	−497	−425	—
Asia			
Borneo			
Pinus caribaea	302	250	92

TABLE 6-21. **High-Lumber-Production-Cost Scenario; Internal Rate of Return; 1979 Constant Prices in Perpetuity (Base Stumpage Prices for Pulpwood)**
(percent)

		Regime	
Region/species	Pulpwood	Integrated, with standard-quality sawtimber	Integrated, with lower-quality sawtimber
North America			
U.S. South			
Pinus taeda, avg.-yield site	12.02	11.70	—
Pinus taeda, high-yield site	13.91	13.40	—
Pacific Northwest			
Pseudotsuga menziesii, avg.-yield site	7.11	5.83	—
Pseudotsuga menziesii, high-yield site	8.85	8.16	—
South America			
Brazil, Amazonia			
Pinus caribaea	17.89	18.64	17.12
Gmelina spp.	27.53	18.58	—
Brazil, Central			
Eucalyptus spp.	20.16	11.08	—
Brazil, Southern			
Pinus taeda	15.57	16.41	15.47
Chile			
Pinus radiata	23.39	15.15	12.48
Oceania			
Australia			
Pinus radiata	10.68	8.06	5.77
New Zealand			
Pinus radiata	11.90	9.55	5.36
Africa			
South Africa			
Pinus patula	19.34	15.48	13.26
Gambia-Senegal			
Gmelina spp.	18.42	15.17	—
Eucalyptus spp.	12.16	10.49	—
Europe			
Nordic			
Picea abies	4.61	5.17	—
Asia			
Borneo			
Pinus caribaea	12.94	12.82	11.21

likely to appreciate in price more rapidly than other sawtimber. Comparing again tables 6-22 and 6-25 provides estimates of the additional returns that would accrue to sawtimber that was perceived by the market to be of sufficiently high quality to justify a 1 percent per annum more rapid price increase. This effect could substantially enhance the competitive position of traditional producers such as North America and the Nordic region if, as is sometimes maintained, the traditional temperate climate regions produce an appreciably higher-quality sawlog.

For example, in this scenario the PNVs of all of the North American and Nordic regions examined perform well, and the better cases exceed the highest returns of plantations in the tropics and South America. A second effect would be to enhance the returns to all integrated regimes vis à vis pulpwood regimes.

Summary of Positive-to-Negative PNV Shifts

This chapter has focused on the sensitivity of the PNVs of the representative plantations to two important types of changes: direct changes in final product prices and changes in processing, distribution, and international transport costs that alter implicit stumpage prices. In chapter 5, the sensitivity of the PNVs to discount rate changes was examined. A necessary (but not sufficient) condition for an industrial plantation investment to be economically viable is that the PNV be positive. Following is a summary

TABLE 6-22. Real-Stumpage-Price Growth Rate 1 Percent, Base Case; Present Net Value, 5 Percent Discount Rate
(1979 U.S. $/hectare)

		Regime	
Region/species	Pulpwood	Integrated, with standard-quality sawtimber	Integrated, with lower-quality sawtimber
North America			
U.S. South			
Pinus taeda, avg.-yield site	2,754	3,931	—
Pinus taeda, high-yield site	4,336	5,821	—
Pacific Northwest			
Pseudotsuga menziesii, avg.-yield site	1,362	2,086	—
Pseudotsuga menziesii, high-yield site	1,963	4,015	—
South America			
Brazil, Amazonia			
Pinus caribaea	4,587	5,855	4,897
Gmelina spp.	3,631	4,561	—
Brazil, Central			
Eucalyptus spp.	4,986	6,059	—
Brazil, Southern			
Pinus taeda	5,466	6,865	5,800
Chile			
Pinus radiata	5,217	6,747	4,833
Oceania			
Australia			
Pinus radiata	3,267	3,544	2,404
New Zealand			
Pinus radiata	4,608	6,292	4,109
Africa			
South Africa			
Pinus patula	4,402	5,512	4,160
Gambia-Senegal			
Gmelina spp.	3,242	3,752	—
Eucalyptus spp.	2,917	4,780	—
Europe			
Nordic			
Picea abies	277	638	—
Asia			
Borneo			
Pinus caribaea	2,953	3,595	2,872

TABLE 6-23. Real-Stumpage-Price Growth Rate 1 Percent, Base Case; Present Net Value, 10 Percent Discount Rate
(1979 U.S. $/hectare)

		Regime	
Region/species	Pulpwood	Integrated, with standard-quality sawtimber	Integrated, with lower-quality sawtimber
North America			
U.S. South			
Pinus taeda, avg.-yield site	361	495	—
Pinus taeda, high-yield site	705	870	—
Pacific Northwest			
Pseudotsuga menziesii, avg.-yield site	−195	−250	—
Pseudotsuga menziesii, high-yield site	14	156	—
South America			
Brazil, Amazonia			
Pinus caribaea	1,174	1,616	1,322
Gmelina spp.	1,196	1,355	—
Brazil, Central			
Eucalyptus spp.	1,493	1,349	—
Brazil, Southern			
Pinus taeda	1,271	1,717	1,430
Chile			
Pinus radiata	1,316	1,278	874
Oceania			
Australia			
Pinus radiata	323	243	22
New Zealand			
Pinus radiata	681	1,066	496
Africa			
South Africa			
Pinus patula	1,128	1,171	846
Gambia-Senegal			
Gmelina spp.	893	927	—
Eucalyptus spp.	527	1,114	—
Europe			
Nordic			
Picea abies	−467	−372	—
Asia			
Borneo			
Pinus caribaea	531	757	559

of the relative sensitivity of the PNVs to changes in the various parameters, that is, the frequency with which PNVs change from a positive to a negative value.

Table 6-28 indicates the number of cases in which a sign change occurs as a result of an increase in the discount rate. The sixteen base cases each of pulpwood and integrated (with standard-quality sawtimber) plantation are examined, as well as the seven variant cases in which the integrated plantations with lower-quality sawtimber were assumed to sell their lumber at a discount of 10 percent in the world market. For the base condition, fifteen of the sixteen pulpwood plantations, all of the sixteen integrated plantations, and all of the seven price-discounting integrated plantations generate positive PNVs using a 5 percent discount rate.

Increasing the discount rate to 10 percent changes the sign of the PNV for two pulpwood plantations, three integrated plantations, and one price-discounting integrated plantation.

Tables 6-29 and 6-30 summarize the shift of PNV from a positive to a negative value in response to the various transport, processing, and distributions cost changes discussed using both 5 and 10 percent discount rates. As with the 5 percent discount rate (table 6-29), very few plantations shift from positive to negative PNVs. The notable ex-

TABLE 6-24. Real-Stumpage-Price Growth Rate 1 Percent, Base Case; Internal Rate of Return
(percent)

	Regime		
Region/species	Pulpwood	Integrated, with standard-quality sawtimber	Integrated, with lower-quality sawtimber
North America			
U.S. South			
Pinus taeda, avg.-yield site	13.23	13.64	—
Pinus taeda, high-yield site	15.13	15.32	—
Pacific Northwest			
Pseudotsuga menziesii, avg.-yield site	8.34	8.29	—
Pseudotsuga menziesii, high-yield site	10.12	10.87	—
South America			
Brazil, Amazonia			
Pinus caribaea	19.40	21.87	20.70
Gmelina spp.	29.27	25.12	—
Brazil, Central			
Eucalyptus spp.	21.60	16.93	—
Brazil, Southern			
Pinus taeda	17.05	18.87	18.14
Chile			
Pinus radiata	24.68	18.72	17.21
Oceania			
Australia			
Pinus radiata	11.90	11.23	10.13
New Zealand			
Pinus radiata	13.22	14.35	12.39
Africa			
South Africa			
Pinus patula	20.71	19.00	17.58
Gambia-Senegal			
Gmelina spp.	19.69	18.76	—
Eucalyptus spp.	13.53	15.95	—
Europe			
Nordic			
Picea abies	5.76	6.66	—
Asia			
Borneo			
Pinus caribaea	14.42	16.13	15.04

TABLE 6-25. Real-Stumpage-Price Growth Rates of 1 Percent for Pulpwood; 2 Percent for Sawtimber, Base Case; Present Net Value, 5 Percent Discount Rate
(1979 U.S. $/hectare)

		Regime	
Region/species	Pulpwood	Integrated, with standard-quality sawtimber	Integrated, with lower-quality sawtimber
North America			
U.S. South			
Pinus taeda, avg.-yield site	2,754	5,902	—
Pinus taeda, high-yield site	4,336	8,628	—
Pacific Northwest			
Pseudotsuga menziesii, avg.-yield site	1,362	4,233	—
Pseudotsuga menziesii, high-yield site	1,963	6,811	—
South America			
Brazil, Amazonia			
Pinus caribaea	4,587	7,655	6,265
Gmelina spp.	3,631	6,154	—
Brazil, Central			
Eucalyptus spp.	4,986	8,872	—
Brazil, Southern			
Pinus taeda	5,466	9,031	7,448
Chile			
Pinus radiata	5,217	10,234	7,255
Oceania			
Australia			
Pinus radiata	3,267	5,550	3,748
New Zealand			
Pinus radiata	4,608	9,527	6,261
Africa			
South Africa			
Pinus patula	4,402	8,066	6,013
Gambia-Senegal			
Gmelina spp.	3,242	5,039	—
Eucalyptus spp.	2,917	6,714	—
Europe			
Nordic			
Picea abies	277	1,410	—
Asia			
Borneo			
Pinus caribaea	2,953	4,890	3,820

ception is for the 20 percent increase in pulpwood processing costs, which shifts seven of the fifteen positive cases to negative, and also one of the sixteen integrated cases.

At the 10 percent discount rate in the base case (see table 6-28), thirteen of both the pulpwood and integrated regimes have positive PNVs. Again, as shown in table 6-30, the sensitivity experiments have only a modest effect in changing the sign of the PNV except for the higher pulpwood processing costs. Here all thirteen of the plantations with positive PNVs for the base case pulpwood regimes become negative. For the thirteen integrated plantations with positive PNVs for the base case, seven PNVs become negative when experiencing the 20 percent higher wood pulp processing costs.

Implications of the Sensitivity Analysis

The sensitivity analysis demonstrates the following: First, the discount rate used is very important, both as a determinant of the size and sign of the PNV and also in the evaluation of the alternative pulpwood and integrated regimes. A higher discount rate tends to favor the shorter-rotation, pulpwood regime. Second, for most of the cost scenarios examined, the sign and size of the PNVs of the thirty-two cases examined are not significantly affected by a change in the sensitivity component. Third, an important exception to the above is found in the wood

TABLE 6-26. Real-Stumpage-Price Growth Rates of 1 Percent for Pulpwood, 2 Percent for Sawtimber, Base Case; Present Net Value, 10 Percent Discount Rate
(1979 U.S. $/hectare)

Region/species	Regime		
	Pulpwood	Integrated, with standard-quality sawtimber	Integrated, with lower-quality sawtimber
North America			
U.S. South			
Pinus taeda, avg.-yield site	361	705	—
Pinus taeda, high-yield site	705	1,172	—
Pacific Northwest			
Pseudotsuga menziesii, avg.-yield site	−195	−107	—
Pseudotsuga menziesii, high-yield site	14	421	—
South America			
Brazil, Amazonia			
Pinus caribaea	1,174	1,901	1,539
Gmelina spp.	1,196	1,616	—
Brazil, Central			
Eucalyptus spp.	1,493	1,783	—
Brazil, Southern			
Pinus taeda	1,271	2,040	1,676
Chile			
Pinus radiata	1,316	1,721	1,181
Oceania			
Australia			
Pinus radiata	323	484	181
New Zealand			
Pinus radiata	681	1,531	803
Africa			
South Africa			
Pinus patula	1,128	1,525	1,102
Gambia-Senegal			
Gmelina spp.	893	1,123	—
Eucalyptus spp.	527	1,437	—
Europe			
Nordic			
Picea abies	−467	−341	—
Asia			
Borneo			
Pinus caribaea	531	952	702

TABLE 6-27. Real-Stumpage-Price Growth Rates of 1 Percent for Pulpwood; 2 Percent for Sawtimber, Base Case; Internal Rate of Return
(percent)

	Regime		
Region/species	Pulpwood	Integrated, with standard-quality sawtimber	Integrated, with lower-quality sawtimber
North America			
U.S. South			
Pinus taeda, avg.-yield site	13.23	14.36	—
Pinus taeda, high-yield site	15.13	16.00	—
Pacific Northwest			
Pseudotsuga menziesii, avg.-yield site	8.34	9.42	—
Pseudotsuga menziesii, high-yield site	10.12	11.91	—
South America			
Brazil, Amazonia			
Pinus caribaea	19.40	22.60	21.34
Gmelina spp.	29.27	26.08	—
Brazil, Central			
Eucalyptus spp.	21.60	17.95	—
Brazil, Southern			
Pinus taeda	17.05	19.46	18.65
Chile			
Pinus radiata	24.68	19.73	18.16
Oceania			
Australia			
Pinus radiata	11.90	12.09	10.93
New Zealand			
Pinus radiata	13.22	15.39	13.37
Africa			
South Africa			
Pinus patula	20.71	19.97	18.49
Gambia-Senegal			
Gmelina spp.	19.69	19.45	—
Eucalyptus spp.	13.53	16.87	—
Europe			
Nordic			
Picea abies	5.76	7.41	—
Asia			
Borneo			
Pinus caribaea	14.42	16.88	15.71

pulp processing costs. It was seen that a 20 percent increase in these costs can have a dramatic effect by reducing the PNV and causing many of the signs to become negative. Fourth, the assumption of a 1 percent growth rate in stumpage prices through time improves the returns and hence the PNVs and IRRs for all cases. Fifth, the assumption of a 2 percent growth rate for sawtimber stumpage of the Northern Hemisphere forests improves dramatically the PNVs of the traditional plantations, giving the high-yield-site North American plantations the highest PNVs observed in the study.

In addition, the analysis demonstrates that the relative performance of the regional plantations is largely invariant to the choice of investment criteria, that is, PNV or IRR,

TABLE 6-28. Present Net Value Sensitivity to Discount Rate

Number of Cases	Pulpwood regime	Integrated regime	Integrated regime with discounted price
Total	16	16	7
Positive PNV (base case using 5 percent discount)	15	16	7
Positive PNV (base case using 10 percent discount)	13	13	6

ical and temperate Southern Hemisphere plantations outperformed those of Europe and North America except for the U.S. South. The best performers of the nontraditional regions are South America and Oceania, particularly New Zealand. However, the African and Asian plantations also consistently perform well.

The best-performing traditional producing region is the U.S. South. In general, the South's performance is more similar to that of the average nontraditional regions than to the performance of the other traditional representative plantations examined. The poorest-performing region is consistently the Nordic region.

of the discount rate, or of the changes in the various cost components. The major exception is found in the effect of higher wood pulp processing costs on the performance of all cases, but especially on that of tropical hardwood pulpwood plantations. Aside from that important exception, however, the trop-

Reference

Fenton, R. T. 1972. "Implications of Radiata Pine Afforestation Studies," *New Zealand Journal of Forest Science* vol. 2, no. 3, pp. 378–388.

TABLE 6-29. Number of Plantations that Shift from Positive Value in Base Case PNV as a Result of a Change in Sensitivity Component (5 Percent Discount Rate)

Sensitivity component	Pulpwood regime	Integrated regime
International transport cost plus 20 percent (effect all on stumpage prices)	0 of 15	0 of 16
International transport cost plus 20 percent (effect shown by stumpage and final market price)	0 of 15	0 of 16
Harvest cost plus 20 percent	1 of 15	0 of 16
Pulpwood processing plus 20 percent	7 of 15	1 of 16
Lumber processing plus 20 percent	none	1 of 16

TABLE 6-30. Number of Plantations that Shift from Positive Value in Base Case to Negative PNV as a Result of Change in Sensitivity Component (10 Percent Discount Rate)

Sensitivity component	Pulpwood regime	Integrated regime
International transport cost plus 20 percent (effect shared by stumpage and final market price)	0 of 13	1 of 13
International transport cost plus 20 percent (effect all on stumpage prices)	1 of 13	1 of 13
Harvest cost plus 20 percent	2 of 13	1 of 13
Pulpwood processing plus 20 percent	13 of 13	7 of 13
Lumber processing plus 20 percent	0 of 13	2 of 13

7

Implications of the Results

Before considering the implications of the base case and sensitivity analysis results, let us briefly review those results. The results of the formal model, reported in chapter 5, indicate that, using a 5 percent discount rate and applying the present net value (PNV) criterion, all of the regions generate positive PNVs for their integrated (sawtimber and pulpwood) plantations, and that all but one generate positive PNVs for their pulpwood regimes. In general, the tropical and temperate Southern Hemisphere regions outperform the traditional, temperate-climate wood-producing regions, with the Nordic region plantation having the lowest PNVs and IRRs (internal rates of return). The U.S. South compared favorably with most tropical and Southern Hemisphere plantations. Increasing the discount rate to 10 percent, of course, diminished the PNV. Using the 10 percent discount rate criterion, all of the northern temperate-climate plantations examined generated negative PNVs except those of the U.S. South. In general, the tropical and temperate Southern Hemisphere plantations continued to generate positive PNVs even at the higher discount rate.

The sensitivity analysis revealed that a 20 percent increase in most of the model's individual cost components was not sufficient to change the sign of the PNV except for marginal cases. In most cases the plantations with the higher PNVs were somewhat more sensitive to increased costs.

An exception occurred, however, when the 20 percent cost increase was also large in absolute terms as in the case of an increase in wood pulp processing costs. In this case the effect upon the PNV was dramatic. This was particularly true for the plantations in which the rapid rate of biological growth combined with somewhat lower stumpage prices. Hence many of the plantations of the tropics, and particularly the hardwood plantations, experienced a radical reduction in their PNV, which often became negative even at the 5 percent discount rate. In many of these cases a 20 percent increase in wood pulp processing costs radically reduced the value of the PNV, often generating a large negative value.

The consideration of development costs, land acquisition costs, and risk can substantially modify the results obtained from the formal model. Specifically, large costs associated with all or any of these factors, or a high risk factor, or both, would militate against the high economic returns that may have been generated by the formal model. However, one of the uses of the formal model is that of estimating the level of development, land acquisition, and other costs that might be incurred without destroying the economic viability of the investment.

The results of this study indicate that the economics of timber growing appear favorable under reasonable assumptions in all of the twelve regions investigated if the risk and development costs discussed in chapter 4 are small. The results of the sensitivity analysis, which tested the robustness of the conclusions of the formal model under a variety of alternative assumptions about costs, were largely invariant to the changing-cost assumptions. An exception to this generalization, however, was found in the results for increased wood pulp processing costs; these results were quite sensitive to higher costs.

Implications of the Results

Comparative and Absolute Advantage

The concepts of absolute advantage and comparative advantage are discussed in chapter 4. As was stated, absolute advantage in the context of this study relates to the ability of the various regions to produce timber and to their performance, relative to other regions, in producing timber. The concept of comparative advantage, by contrast, relates not only to the ability of land within a region to produce timber, but also to its productivity in plantation forestry relative to alternative economic activities, such as agriculture. As such, comparative advantage deals with the choice of the "best" economic activity in which to engage the land, that is, the use that offers the highest economic return. The estimates of relatively high PNVs for the representative plantations studied, plus the fact that the data used in the study are drawn largely from the experience of forest plantations that have displaced other land uses or are on formerly idle land, suggest that plantations are the highest-value use of the land, or the use with the comparative advantage.

Long-Term Supply Potential of Plantations

As indicated in chapter 1, volumes of future production from industrial forest plantations could be great. In order for plantations to have sizable impact upon global long-term supply and thus upon world markets, the two conditions of reasonably favorable underlying economics and substantial land areas available for production must be present. As was indicated, high-yielding forest plantations are capable of producing 15 to 20 cubic meters per hectare per year, so that a modest 140 million hectares, or 5.0 percent of the world's forest land capable of yielding 10 cubic meters per year could have met world demand in 1978.

Although a complete survey has not been made, it is clear that substantial areas of land have the biological potential to support high-yielding forest plantations. Land with high-yielding potential is found throughout the regions examined and includes, importantly, almost 20 million hectares within the United States that have the biological potential for high-yielding plantations.[1] Thus, although no systematic attempt is made to estimate the long-term output potential of high-yielding plantations, prima facie evidence suggests that the worldwide biological potential is great,

---

[1]This is the productivity class that produces 120 or more cubic feet per acre in fully stocked *natural* stands.

and the economics also appear to be favorable in many regions, given recent prices and costs.

Implications for Forest Resources

The study examines the industrial forest plantation within the context of a view that perceives a growing role for plantation forests. The implications of this mode of resource production are potentially profound. Just as modern agriculture involves decisions as to location, crop type, technological inputs, and management mode, so, too, forest decisions increasingly involve questions of forest plantation location, species choice, technological inputs, and management regime. The locational question in particular deserves attention. With the advent of the plantation, forest location became a decision variable. And the decision is not simply confined to one region or country, but now location can involve, in principle, any region of the world. Thus, traditional, temperate forest-producing regions of the Northern Hemisphere now compete with regions of the tropics and temperate Southern Hemisphere, some sites of which have never been forested. Species is also a decision variable. The tradeoffs between high biological growth, advantageous location vis-à-vis major markets, development costs particularly in primitive regions, and both biological and political risks now become very real.

More importantly, the favorable economics of plantations in the tropics and temperate Southern Hemisphere, as well as in the U.S. South, suggest the southward shift of the world's forest resource production. This trend can be expected to be accentuated as the old-growth temperate forests continue to be depleted. Although much of the production of the Southern Hemisphere plantations will be consumed locally, large volumes should also find their way to the major world markets of Europe, the northeastern United States, and Japan.

Implications for U.S. Forestry

Within this context it would be expected that the United States, and particularly the U.S. South, will continue to be a major producer of industrial wood. The formal results of the study suggest that the economics of plantation forests in the United States are favorable. These basic quantitative findings are reinforced by consideration of development costs and risk. A well-developed infrastructure in the United States implies low development costs, while political stability together with biological experience both suggest low levels of risk associated with the development of forest plantations in the United States relative to the other geographic locations examined.

More generally, as has been mentioned, the findings of this study suggest a shifting of worldwide production of industrial wood southward. Biological and economic considerations favor locations in the tropics and the temperate Southern Hemisphere even though for many locations advantages might be offset to some degree by considerations of development costs and risk (see chapter 8 for discussion of biological considerations).

Implications for the Structure of World Production and Trade

While the principal objective of this study is an examination of the economic potential for forest plantations for various regions around the globe, the analysis does provide some insights as to which consuming regions are likely to be the major markets for the various producing regions. In general, the South American and African plantations are likely to find Europe the dominant market for their exports, while Oceanic and Southeast Asian exports flow to Japan. North American plantations will find their major markets in the United States and will compete for foreign markets in both Europe and Japan.

Also, one would expect the current rather

modest net forest resource trade flows to the Southern Hemisphere from the Northern Hemisphere temperate-climate regions to be replaced by growing forest resource net trade flows originating in the Southern Hemisphere and terminating in the major markets of the Northern Hemisphere. One might view the current net trade flows going to northern markets from New Zealand and Chile as the fledgling stage of what may well become substantial flows by the early part of the twenty-first century.

8

Ecological Implications of Tropical Plantation Forestry

Introduction

Although opinions differ on the rate of tropical deforestation and on the number of decades of survival remaining for tropical forests (World Bank, 1978; Lanly and Clement, 1979; Myers, 1980; Persson, 1974; Pringle, 1976; Richards, 1973; Sommer, 1976; Spears, 1979; Zerbe and coauthors, 1980), no one denies the ongoing destruction of the world's tropical forests. In view of the extensive deforestation of parts of the tropics and the pessimistic prognosis for the remaining tropical forests, it is clear that plantation forestry, including agroforestry, which intermixes tree growing and cropping, must play a rapidly expanding role in the production of forest products in most tropical countries. According to John Spears (1978) of the World Bank, by the year 2000 the developing world will need a minimum of 20 to 25 million hectares of new plantations just

---

The author of this chapter is Gary S. Hartshorn, of the Tropical Science Center, San Jose, Costa Rica. Editing assistance by Catherine S. Tunis, Research Assistant, Resources for the Future.

for fuelwood consumption. Yet, continuation of the current rate of plantation establishment will produce less than one-tenth of the projected requirement. The firewood crisis affecting the drier tropics like sub-Saharan Africa has been well publicized (e.g., Eckholm, 1975). Less well known is the rapid deforestation of once richly forested countries such as Thailand, the Philippines, Malaysia, Ivory Coast, and Costa Rica. Depletion of the species-rich lowland dipterocarp forests of peninsular Malaysia has been so rapid, that government foresters estimate Malaysia will be a net importer of timber by 1990.

Even in countries undergoing moderate or slow loss of tropical forests, plantation forestry and agroforestry can play important roles in reducing pressures on remaining natural forests as well as alleviating local scarcities of wood. Decision makers are realizing that unless energy and wood needs of local people are met, the efforts to conserve forests will be doomed. Easily accessible (that is, exploitable) forests no longer exist in most tropical countries; hence populated areas may experience scarcity of wood even though extensive areas of forests remain in remote regions of a country. For example, the Andean Altiplano and intermountain valleys of Ecuador, Peru, and Bolivia have been heavily populated and deforested for centuries, contrasting starkly to the vast forests covering their Amazonian slopes and lowlands (e.g., Freeman and coauthors, 1980).

Fast-growing trees have the potential to meet the increasing demands for fuelwood, paper pulp, lumber, and site rehabilitation in developing tropical countries. Decision makers must understand the ecological implications of and constraints on tropical plantation forestry if ecological failures are to be avoided and if the potential of tropical trees is to be successfully used. Ecological aspects of appropriate land use, site rehabilitation and degradation, pests and diseases, and plantation species, are reviewed in this chapter from a tropical perspective. A commentary on the role of plantations in the issue of tropical deforestation is also offered.

Appropriate Land Use

Trees grow in a great variety of environmental sites and conditions, and indeed have been and can be successfully introduced into areas historically devoid of trees. Tree introduction and reforestation have some outstanding successes, for example, *Eucalyptus globulus* in the high tropical Andes, *Tectona grandis* in the tropical Far East, and *Pinus radiata* in Chile and New Zealand. Reforestation and afforestation are certainly not appropriate for all tropical areas. Land capable of sustained production of crops or pasture seldom can be justified for use in plantation forestry. At the other extreme, extensive areas of rugged topography with high rainfall need to be protected by natural vegetation so that streamflow from watersheds is regulated, thereby lessening streamflow extremes, and so that soil erosion and streambed aggradation are minimized (Pereira, 1973).

Between the extremes of land suitable for sustained intensive agriculture and of watersheds requiring continuous forest cover for protection is the great bulk of tropical land suitable for pasture, semipermanent crops, or forestry. These intermediate forest lands in humid (for example, a tropical moist forest), perhumid (for example, a tropical wet forest), and superhumid (for example, a tropical rainforest) ecological life zones (Holdridge, 1967) have been traditionally used by "slash-and-burn" agriculturalists, generally through shifting cultivation (cf. Braun, 1974; Spencer, 1977; and Watters, 1971). Under low population densities, shifting cultivation is an ecologically sound use of tropical forests in areas unsuitable for sustained agriculture because of poor soils, steep terrain, and high rainfall. Third World human population explosion and unequal distribution of arable land force increasing numbers of peasants to practice slash-and-burn agriculture on less suitable land in the wet lowlands or higher in the watersheds. Larger clearings and shortened fallow periods—practices often used by inexperienced colonists from different ecological regions—lead to site degradation.

The quickening advance of the "agricultural frontier" into the wet tropics has caused extensive deterioration of the natural resource, including destruction of timber, lowered soil fertility, increased soil erosion, and loss of native habitat (Hecht, 1980; Nations and Nigh, 1978; and Smith, 1976). Heterogeneous tropical forests are the most species-rich ecosystems on earth, with an astonishing array of highly evolved interactions and interdependence of species. The conversion of forest to nonforest, that is, loss of habitat, is devastating to most species, and may lead to the extinction of native species (Eckholm, 1978; Myers, 1979). Native species loss is often due to the restricted geographic range of many species inhabiting tropical forests (Prance, 1981). Wherever roads are built, opening up "new land" for oil exploration or logging, slash-and-burn agriculturalists soon colonize the road margins—or even precede the road builders. In fact the dominant landscape along most roads in the western tropics is not productive crops or forest, but scrubby, successional vegetation or brushy pasture.

The abandoned agricultural lands and unproductive pastures prevalent in the tropics are generally biologically suitable for plantation forestry. Land use capability studies indicate large areas in the perhumid tropics where production forestry is the most appropriate and sustainable land use without causing serious, irreversible environmental degradation (Hartshorn, 1977; Tosi, 1976). Unfortunately, decision makers in developing countries largely ignore soil capability, preferring instead politically expedient actions such as colonization or agribusiness.

Tree plantations are not suitable for every site or deforested area. Soil capability studies are an essential prerequisite for determining general suitability for plantation forestry. Detailed knowledge of site capability is just as important as information on appropriate plantation species.

Site Rehabilitation

If trees can be established on a degraded site, they usually contribute to site rehabilitation. Tree plantations generally improve soil structure, increase soil organic matter and fertility, lessen erosion, favorably change microclimate, and improve local habitat. Tree plantations stabilize degraded land that otherwise would continue to deteriorate. Although practically any degraded site can be improved with tree plantations, such ecological successes may be very costly or even uneconomic, particularly if the primary objective is wood or fiber production. Nevertheless, successful afforestation of savanna or grasslands, as in Africa, Brazil, and the Venezuelan llanos, has produced quite acceptable yields of wood.

Establishment of trees may be the most difficult aspect of site reclamation. Protection from grazing animals, fire, pests, and weedy competition is often necessary to the establishment of tree plantations. In severely deforested areas such as the African Sahel, the local demand for fuelwood caused premature harvesting of planted trees, that is, harvesting before substantial growth could occur. The desperate need for fuelwood in many parts of the tropics (Eckholm, 1975; NAS, 1980) and the lack of adequate appreciation of trees in the forested tropics clearly indicate the importance of considering native people's needs and attitudes in assessing the risks and problems associated with tropical plantation forestry. General governmental disinterest or incompetence in the protection of forests and plantations from illegal or undesirable human interventions (for example, squatters and clandestine tree cutters) further complicates plantation forestry efforts in tropical, developing countries.

Rehabilitation of degraded sites is most critical in watersheds where slash-and-burn agriculture is practiced. A country's important watersheds tend to be in the highest rainfall areas (perhumid or superhumid ecological life zones); hence land use for seasonal or annual crops is particularly inappropriate for areas usually suitable only for protection or production forests.

Tropical countries have until recently ignored protection and management of watersheds above hydroelectric and irrigation reservoirs. The destruction of natural vegetation by slash-and-burn agriculturalists causes a severalfold increase in erosion, drastically reducing the capacity and therefore the life of the reservoir. Project development for major tropical impoundments seems to assign a standard life of fifty years for a reservoir; however, rapid accumulation of sediments often reduces the useful period to less than half the original estimate, as has occurred with reservoirs in El Salvador, Colombia, and the Dominican Republic. Deforestation and inappropriate land use in the Panama Canal watershed has caused considerable sedimentation in Lake Alajuela and is seriously threatening the dry-season water supply for operation of the Panama Canal (Wadsworth, 1978). Rehabilitation efforts have begun in this watershed through the

Panama Office for Renewable Natural Resources (RENARE) with funding from the U.S. Agency for International Development and the Panamanian government. Reforestation of the watershed seemed to be the obvious solution in this situation, but consideration of the 60,000 people living there points to the need for a combination of reforestation and incentives to encourage appropriate agricultural practices and natural resource conservation (Bob Otto, AID, personal communication, June 2, 1981).

Plantation forestry can play an important role in helping to protect threatened watersheds. But in areas where harvesting activities may be too damaging to the watershed, natural regeneration techniques may be more appropriate.

Site Degradation

Even though most forestry plantations are established on degraded or nonforest lands and generally contribute to site reclamation, there is legitimate concern that plantation forestry may cause certain types of ecological deterioration. Specific issues include loss of soil fertility with reduced productivity in successive rotations, detrimental effects of site preparation and tree harvesting on soils, and reduction in water yields in dry areas.

Most inhabitants of developed countries have an image of the tropical forests as lush, green, fertile jungle. Indeed, the aboveground luxuriance of many tropical forests would suggest that the soil below is bounteous and fertile. For the majority of tropical forest soils nothing could be further from the truth; many (for example, Oxisols, Ultisols, and Spodosols) have high erodibility and low-nutrient fertility. The latter condition is due to low organic-matter content, low cation-exchange capacity, low base saturation, and a predominance of 1:1-type clays (Sanchez, 1976). In contrast to 2:1- and 2:2-type clays, 1:1-type clays have appreciably less adsorptive capacity, hence colloidal organic matter is the primary source of nutrients. Some of the nutrient-poor Oxisols and Spodosols have a surprisingly thick humus layer on the soil surface. Fertile agricultural soils are also prevalent, of course, such as those derived from andesitic volcanic ash (Andepts) in Central America, East Africa, and Java, as well as the *terra roxa* soils (Alfisols) of Brazil.

The luxuriance of tropical forests is built upon and maintained by a highly evolved and remarkably efficient nutrient-cycling system. Keystone to the system is a symbiotic association of tree roots and mycorrhizal fungi that quickly absorb nutrients, preventing substantive leakage of nutrients from the ecosystem (Herrera and coauthors, 1978; Went and Stark, 1968). Although mycorrhizal inoculation of pine seedlings is a well-established nursery practice, recent evidence indicates mycorrhizal fungi are necessary to the survival and growth of most tropical trees (Janos, 1975); hence mycorrhizae may be a key factor to the establishment of tree plantations on nonforest or degraded lands.

Site degradation through loss of fertility may have varied and complex causes, including high nutrient demand by the plantation species, nutrient export in harvested wood, and inappropriate site preparation or harvesting. Evidence of a tropical forest plantation depleting or requiring specific nutrients is meager (e.g., Kadeba, 1978). However, it is not unreasonable to assume that a single-species plantation will differentially extract specific nutrients, similar to many short-term agricultural crops. Successive rotations of the same tree species may cause imbalance in the small nutrient pool in the soil, leading to nutrient-limited growth.

The growing trend toward complete utilization of wood or fiber resources in forests and plantations has prompted questions about the ecological consequences of total harvesting. In addition to removal of forestlike structure and abrupt exposure of the soil, complete harvesting of mixed, indigenous tropical hardwoods by clear-cutting results in a major export of wood-stored nutrients from the site (Ewel and Conde, 1978). Similar nutrient depletion may occur when tropical sites planted with pine, eucalyptus, or other fast-growing pioneer species are harvested, but data on these species are sparse. Harvesting of all boles and branches may seriously deplete the low nutrient stocks of many tropical soils to the extent that the next harvest is delayed or impossible. Serious site degradation to the point of precluding a subsequent harvest is most probable on the virtually sterile white-sand soils (Spodosols). As part of an international project in San Carolos, Venezuela, German scientists planted rubber trees (*Hevea brasiliensis*) after clearing the virgin forest on white-sand soils, but very few rubber trees survived on the extremely poor site (Hartshorn, personal observation).

Sandy tropical soils, especially Ultisols and Spodosols, are highly susceptible to surface erosion. Removal of the vegetative cover, whether natural forest or plantations, exposes the unprotected surface soil to the awesome erosive capability of tropical rains. There is no question that logging increases erosion, but this is very poorly documented (for example, see Anderson, 1972; Pereira, 1973; Sanchez, 1976).

Mechanical clearing by bulldozers and harvesting with tracked vehicles or skidders generally cause serious damage to the soil. Bulldozing of forest or brush into windrows may remove most of the fertile topsoil from some soils, such as occurred in the initial clearings for gmelina establishment on sandy soils at the Jari plantation in Brazil. Physical compaction by heavy machinery may substantially reduce the infiltration capacity of the soil, leading to greater surface runoff and erosion. Less damaging site preparation and harvesting techniques, for example, labor-intensive methods, may be more appropriate on sensitive sites. Afforestation of savanna soils may tend to deplete soil moisture because tree roots can tap a deeper volume of soil than do grass roots. In arid regions greater evapotranspirative use of limited soil moisture by trees may be undesirable. The potential drying effect of exotici-tree afforestation in arid regions is considered insignificant when compared with the numerous ecological benefits of afforestation (Cozzo, 1976).

Pests and Diseases

The risk of pests or diseases ruining a plantation is a prominent concern of the forester, company, or government agency considering commercial plantation forestry. This section addresses that concern, with emphasis on the situation for exotic (nonnative) tree species, since exotics are often used in tropical forest plantations.

Tropical plantation forestry has not yet had the centuries of botanical exploration, plant introduction, and economic successes and failures associated with edible plants. Nevertheless, the mixtures of successes and failures found with food plants are occurring in tropical plantation forestry, albeit on a much smaller economic scale. The ubiquitous pine and eucalyptus trees document their successful pantropical dissemination. Only seventeen years ago, a listing of forest tree diseases reported no entries for *Pinus caribaea*, *Eucalyptus deglupta*, and *Swietenia* spp. (IUFRO, 1964).

Although climate and soil factors will affect the success or failure of exotic species, their success is largely attributed to having left behind or escaped "from diseases and pests that have evolved along with them" (Baker, 1970, p. 109). The failure of Henry Ford's vast rubber plantations at Fordlandia in the Brazilian Amazon because of the fungal leaf blight *Dothidella ulei* is often cited as the chief example demonstrating that extensive monocultures in a species' native region are impossible. The mahogany shoot borer *Hypsipyla grandella* has seriously damaged native area plantations of mahogany (*Swietenia macrophylla* and *S. mahogani*, Meliaceae) and Spanish cedar (*Cedrela odorata*, Meliaceae) and has received considerable research effort (Whitmore, 1976a, 1976b), yet most Asian and African Meliaceae, especially *Toona*, are resistant to the tropical American mahogany shoot borer. Tropical American *Cedrela* is commonly planted in Africa, unmolested by local pests.

While exotic species have left behind "diseases and pests that have evolved along with them," they may also be moving to a new location where indigenous pests and diseases can successfully adapt to the exotics as hosts, and for which the exotics have no evolved resistance. Sri Lanka's preeminence as a tea exporter had its origin in the rapid demise of coffee (an exotic) in the eighteenth century because of a leaf spot disease caused by the fungus *Hemileia vastatrix* (Baker, 1970). In Southeast Asia the native *Hypsipyla robusta* attacks the introduced *Swietenia macrophylla* (Chaiglom, 1975), and in the Virgin Islands the introduced *Chukrasia* (from Bangladesh) is attacked by native *H. grandella* (J. L. Whitmore, personal communication, 1978). *Anthocephalus chinensis* (Rubiaceae) from Southeast Asia, a successful plantation tree in the 1960s, suffered complete mortality as a result of fungal attacks on four- to five-year-old stands in Turrialba, Costa Rica, and is no longer even considered as a candidate species for plantation forestry. Leaf-cutter ants (*Atta*) are a major pest to *Gmelina arborea* (an exotic) in tropical America, as well as to many other trees and crops. The heartwood of young *Eucalyptus deglupta* (Myrtaceae) trees is attacked by *Coptotermes crassus* termites in Turrialba, Costa Rica (Ford, 1980). In peninsular Malaysia termites attack live wood of introduced pine trees (Tho Yow Pong, personal communication, 1979).

There is a considerable literature on pests and diseases switching from native hosts to introduced tree species (e.g., Camacho, 1966; Drooz and Bustillo, 1972; Ford, 1980; Gray, 1972; Ivory, 1977; Muchovej and coauthors, 1978; Schonherr, 1977). Reasons for insects switching to exotic hosts are not well understood; however, Gray (1972) concludes that the abundant and extensive resource offered by a monoculture is usually of major importance. Strong (1974) found that the number of insect pests of cacao (*Theobroma cacao, Sterculi aceae*) is best described as a function of the area in cultivation rather than time since introduction. In fact, for similar-sized areas, the number of insect pests did not differ significantly between native and non-native cacao-producing areas. Strong's thesis predicts that pest and disease problems will increase as the area of plantation increases, regardless of the crop's geographic origin. The outbreak of serious diseases on exotic crops in major exporting areas, such as happened with cacao in Ghana (see Baker, 1970) and coffee in Sri Lanka, or now with sugarcane in Cuba, conform to Strong's predictions. The clear implications are that native pests and diseases switch to the plantation crop and that as the area planted increases, more diseases and pests will switch to the increasingly abundant resource.

The appropriateness of site may be an underestimated factor in susceptibility or resistance of a species to pests and diseases. A plantation species may have serious pest or disease problems if grown "off-site," that is, grown under unsuitable soil or site conditions that may put the trees under stress and make them more prone to attack. On the other hand, an excellent site may lessen or eliminate pest or disease problems, for example, on fertile alluvium in Surinam, *Cedrela angustifolia* trees attacked by *Hypsipyla grandella* have better average growth than nonattacked trees in the same plantation (Vega, 1976).

Insect outbreaks in exotic tree plantations occasionally occur during droughts (Brown, 1965b, cited in Gray, 1972). Trees under moisture-stress generally have less resistance to insect attack and may also have a decreased ability to recover.

The biological risks of pests and diseases for tropical plantation forestry are probably no greater than they are in extratropical regions (Johnson, 1976). Tropical plantation forestry depends on appreciable wild populations of tree species with considerable genetic variability, hence tropical foresters and tree geneticists have the opportunity to select for resistant individuals and species. And plantation forestry can fairly easily substitute one species for another, in contrast to the major edible crops which lack substitute species in case of an epidemic. Tropical regions also have the advantage of much shorter rotation periods—fast-growing tropical trees are being harvested for pulp in only six to ten years. The much shorter generation time of tree plantations in the tropics as opposed to the temperate zones lowers the biological risks for tropical plantation forestry. However, the tropical climate also enhances the growth of insects and diseases and the rate at which they compete for food.

Plantation Species

If someone asked the national forest service in almost any tropical country what species to plant commercially, it is highly probable that the recommendations would be drawn from a short list of species. If information on the continent, elevation, and rainfall were available, most tropical foresters could tell you with some certainty the recommended species. Although local exceptions occasionally occur, tropical plantation forestry depends on a rather small set of species. The genera *Pinus* and *Eucalyptus* dominate the list of a few dozen species that have a high potential for successful forest plantations, following the testing of perhaps 1,000 species at several tropical forestry research centers over the past decades. Some 700 tree and shrub species made the master list of potential firewood species (NAS, 1980).

Recommended forest plantation species are those for which successful production and utilization technology exists.

1. They are often pioneer species, that is, they perform well in clear-cut or deforested areas with poor soils.
2. They usually have relatively simple seed collection, storage, and nursery technology.
3. They should be relatively productive, that is, fast-growing.
4. Their wood and other products should have a recognized utility and marketability.

In the future, the potential for a species to be genetically improved will most likely also be an important criterion.

Additional research on potential forest plantation species is needed. If fast-growing tropical trees are to make significant contributions to the urgent needs of developing countries, tropical foresters and forestry agencies must rapidly increase the available knowledge about potential species, matching species to site, on pest and disease susceptibility, genetic selection and improvement, as well as general silviculture and autecology.

The most widely preferred species for commercial forestry plantations in the tropics largely come from moist and dry ecological life zones (Holdridge, 1967). The advance of the agricultural frontier into the wet lowlands will require different species for reforestation. A concerted effort is needed to find the fast-growing tree species that can be used in plantations to meet the increasing needs for wood products and site reclamation in the wet tropics.

Yet the ecological problems and environmental degradation rampant in the tropics cannot await another decade of research. Massive plantation forestry can make a significant impact by supplying tropical and extratropical demands for wood products, as well as minimizing the pervasive ecological deterioration so common in the tropics.

Plantation Forests and Indigenous Forests in the Tropics

Indigenous forests in many areas of the tropics have been and continue to be destroyed or degraded; some of the results are losses of habitat and concurrent losses or threats to native plant and animal species, soil erosion, sedimentation, floods and droughts, desertification, and diminished land productivity. Given the diversity of species and ecosystems in the tropics, and the dearth of information on these, it is impossible to quantify the benefits forgone through the loss of tropical forests.

Tropical forest conversion to nonforest use is primarily due to expansion of the agricultural frontier through slash-and-burn agriculture. Direct conversion of natural forest to plantations is a minuscule percentage of the amount of tropical deforestation. However, direct conversions do occur, sometimes on a large scale (such as, Jari).

Since the social and ecological opportunity costs of establishing tropical forest plantations on the abundance of abandoned or degraded land suitable for plantation forestry would be relatively low (or negative) when compared with the benefits forgone by the destruction of indigenous tropical forests, logic would dictate that conversion of natural forest directly to plantations should be avoided.

There are many reasons for conserving the tropical forest resources, but the pressures for using and renewing these resources are immediate and unavoidable. Research and the development of positive solutions such as forest plantations and agroforestry may be the best way to take some of the pressure off the remaining indigenous forests.

Summary

Rampant deforestation of the tropics requires major expansion of tree plantations to meet the projected demands for forest products. Fast-growing trees have the potential to meet these rapidly increasing demands in developing tropical countries. Serious ecological problems of tropical plantation forestry that need to be avoided include inappropriate land use, site degradation, pests and diseases, and the dangers associated with the limited set of plantation species being promoted.

Rapid population growth and inequitable distribution of arable land in tropical developing countries are forcing slash-and-burn agriculturalists into the wet lowlands and higher in the watersheds, where land is unsuitable for repetitive cropping. Inappropriate land use causes site degradation that can often be remedied through reforestation. Despite several potential site improvements associated with plantation forestry, some site degradation may also occur. Successive rotations of a single-species plantation may differentially deplete certain nutrients, resulting in diminished growth rates. Complete harvesting may impoverish the site through export of nutrients in the harvested wood. Mechanical site preparation or harvesting may greatly increase soil erosion or lead to physical deterioration of the site. Labor-intensive methods may be more appropriate in some cases.

The success of exotic tree species has been in part attributed to escaping from native pests and diseases. Yet there is also danger for exotics in that pests and diseases to which they are susceptible and for which they have no developed resistance may be at their new site. Recent evidence suggests the accumulation of pests and diseases is related to area in plantation. The favorable tropical climate tends to reduce biological risks by reducing rotation ages, but it also increases risks by providing a perfect environment for the growth of insects and diseases.

The dependence upon a small number of tree species for tropical plantations may not be wise or adequate for the reforestation needs of the wet tropics. In spite of the considerable need for research on potential species, site indexing, pest and disease susceptibility, and genetic selection and improvement, the greatly increasing demands for wood products and site rehabilitation demand an urgent and massive increase in tropical forestry plantations. The direct conversion of natural forests to plantations constitutes a minuscule percentage of the total tropical deforestation, but it does contribute to the costs thereof. The establishment of tropical forest plantations on the abundance of suitable lands that have been abandoned by the advancing agricultural frontier has lower social and ecological opportunity costs than has the conversion of indigenous forests to forest plantations. Therefore the conversion of natural tropical forests to plantation forests should be avoided.

References

Anderson, A. 1972. "Devastation on the Amazon?" *Organic Gardening and Farming* pp. 90–93.

Baker, H. G. 1970. *Plants and Civilization* (Belmont, Calif., Wadsworth).

Braun, H. 1974. *Shifting Cultivation in Africa* (FAO, Rome).

Camacho, E. 1966. "Dano que las abejas jicotes del genero *Trigona* causan a los arboles de *Macadamia*," *Turrialba* vol. 16, no. 2, pp. 193–194.

Chaiglom, D. 1975. *Dangerous Insect Pests of Forest Plantations in Thailand.* FAO/IUFRO/DI/75/1-1.

Cozzo, D. 1976. *Tecnologia de la forestacion en Argentina y America Latina*, Hemisferio Sur (Buenos Aires).

Drooz, A. T., and A. E. Bustillo. 1972. "*Glena bisulca*, a Serious Defoliator of *Cupressus lusitanica* in Colombia," *Journal of Economics and Entomology* vol. 65, no. 1, pp. 89–93.

Eckholm, E. P. 1975. *The Other Energy Crisis: Firewood.* Worldwatch Paper No. 1. (Washington, Worldwatch Institute).

———. 1978. *Disappearing Species: The Social Challenge.* Worldwatch Paper No. 22. (Washington, Worldwatch Institute).

Ewel, J., and L. Conde. 1978. "Environmental Implication of Any Species Utilization in the Moist Tropics," in Proceedings of a conference, *Improved Utilization of Tropical Forests* (Madison, Wis., Forest Products Laboratory), pp. 106–123.

Ford, L. B. 1980. *A Survey of Pests in Forest Plantations in Costa Rica.* CATIE, Informe Tec. No. 7 (Turrialba, Costa Rica).

Freeman, P. H. B. Cross, R. D. Flannery, D. A. Harcharik, G. S. Hartshorn, G. Simmonds, and J. D. Wilson. 1980. *Bolivia: State of the Environment and Natural Resources.* (McLean, Va., USAID/JRB Associates).

Gray, B. 1972. "Economic Tropical Forest Entomology," *Annual Revue of Entomology* vol. 17, no. 6030, pp. 313–354.

Harcombe, P. A. 1977. "Nutrient Accumulation by Vegetation During the First Year of Recovery of a Tropical Forest Ecosystem," in J. Cairns, Jr., K. L. Dickson, and E. E. Herricks, eds., *Recovery and Restoration of Damaged Ecosystems* (Charlottesville, University of Virginia Press), pp. 347–378.

Hartshorn, G. S. 1977. *Criterios para la clasificacion de bosques y la determinacion del uso potencial de tierras en Paraguay.* FAO/FO; DP/PAR/72/001, Informe Tec. No. 8 (Asuncion).

Hecht, S. B. 1980. "Some Environmental Effects of Converting Tropical Rainforest to Pasture in Eastern Amazonia." Ph.D. dissertation, University of California, Berkeley.

Herrera, R., C. F. Jordan, H. Klinge, and E. Medina. 1978. "Amazon Ecosystems: Their Structure and Functioning with Particular Emphasis on Nutrients," *Interciencia* vol. 3, no. 4, pp. 223–231.

Holdridge, L. R. 1967. *Life Zone Ecology* (San Jose, Costa Rica, Tropical Science Center).

IUFRO. 1964. *Diseases of Widely Planted Forest Trees* (Oxford).

Ivory, M. H. 1977. "Preliminary Investigations of the Pests of Exotic Forest Trees in Zambia," *Commonwealth Forestry Revue* vol. 56, no. 1, pp. 47–56.

Janos, D. P. 1975. "Effects of Vesicular-Arbuscular Mycorrhizae on Lowland Rainforest Trees," in F. E. Sanders, B. Mosse, and P. B. Tinker, eds., *Endomycorrhizae* (London, Academic Press), pp. 437–446.

Johnson, N. E. 1976. "Biological Opportunities and Fast-Growing Plantations," *Journal of Forestry* vol. 74, no. 4, pp. 206–211.

Kadeba, O. 1978. "Nutritional Aspects of Afforestation with Exotic Tree Species in the Savanna Region of Nigeria," *Commonwealth Forestry Review* vol. 57, no. 3, pp. 191-199.

Lanly, J. P., and J. Clement. 1979. *Present and*

Future Natural Forest and Plantation Areas in the Tropics. FAO/FO:Misc. 79/1 (Rome).

Muchovej, J. J., F. C. Albuquerque, and G. T. Ribeiro. 1978. "*Gmelina arborea*: A New Host of *Ceratocystis fimbriata*," *Plant Disease Reporter* vol. 62, pp. 717–719.

Myers, N. 1979. *The Sinking Ark: A New Look at the Problem of Disappearing Species* (New York, Pergamon Press).

―――. 1980. *Conversion of Tropical Moist Forests*, (Washington. National Research Council, National Academy of Sciences).

National Academy of Sciences. 1980. *Firewood Crops: Shrub and Tree Species for Energy Production* (Washington, National Research Council).

Nations, J. D., and R. B. Nigh. 1978. "Cattle, Cash, Food and Forest: The Destruction of the American Tropics and the Lacandon Maya Alternative," *Culture Agricultural* vol. 6, pp. 1-5.

Pereira, H. C. 1973. *Land Use and Water Resources in Temperate and Tropical Climates* (New York, Cambridge University Press).

Persson, R. 1974. *World Forest Resources*, Research Note 17 (Stockholm, Sweden, Royal College of Forestry).

Prance, Ghillean T., ed. 1981. *Biological Diversification in the Tropics: Proceedings of the Fifth International Symposium of the Association for Tropical Biology*, held at Macuto Beach, Caracas, Venezuela, February 8–13, 1979 (New York, Columbia University Press).

Pringle, S. 1976. "Tropical Moist Forests in World Demand: Supply and Trade," *Unasylva* vol. 28, nos. 112–113, pp. 106–118.

Richards, P. W. 1973. "The Tropical Rainforest," *Scientific American* vol. 229, no. 6, pp. 58–68.

Sanchez, P. A. 1976. *Properties and Management of Soils in the Tropics* (New York, Wiley Interscience).

Schonherr, J. 1977. "Forstentomologie und Forstschutz in Brasilien," *Zeitschrift fur Angewandte Entomologie* vol. 82, no. 3, pp. 284–288.

Smith, N. J. 1976. "The Transamazon Highway: A Cultural and Ecological Analysis of Settlement in the Humid Tropics." Ph.D. dissertation, University of California, Berkeley.

Sommer, A. 1976. "Attempt at an Assessment of the World's Tropical Forests," *Unasylva* vol. 28, nos. 112–113, pp. 5–25.

Spears, J. S. 1978. "Wood as an Energy Source: The Situation in the Developing World." Paper presented at 103rd Annual Meeting, American Forestry Association, Hot Springs, Ark.

―――. 1979. "Can the Wet Tropical Forest Survive?" *Commonwealth Forestry Review*, vol. 58, no. 3, pp. 165–180.

Spencer, J. E. 1977. *Shifting Cultivation in Southeastern Asia* (Berkeley, University of California Press).

Strong, D. R., Jr. 1974. "Rapid Asymptotic Species Accumulation in Phytophagus Insect Communities: The Pests of Cacao," *Science* vol. 185, pp. 1064–1066.

Tosi, J. A. 1976. *La zonificacion ecologica preliminar de la region de Darien en la Republica de Colombia*. (San Jose, Costa Rica, Centro Cientifico Tropical).

Vega, C. L. 1976. Influencia de la silvicultura en el comportamiento de Cedrela en Surinam," in J. L. Whitmore, ed., *Studies on the Shootborer Hypsipyla grandella (Zeller) Lep. Pyralidae*, CATIE, vol. III (Turrialba, Costa Rica), pp. 22–49.

Wadsworth, F. H. 1978. *Deforestation—Death to the Panama Canal*. Proceedings of the U.S. Strategy Conference on Tropical Deforestation (Washington).

Watters, R. F. 1971. *Shifting Cultivation in Latin America* (Rome, FAO).

Went, F. W., and N. Stark. 1968. "Mycorrhiza," *Bioscience* vol. 18, pp. 1035–1039.

Whitmore, J. L., ed. 1976a, b. *Studies on the shootborer Hypsipyla grandella (Zeller) Lep. Pyralidae*, CATIE, vol. II; vol. III (Turrialba, Costa Rica).

World Bank. 1978. Forestry Sector Policy Paper (Washington).

Zerbe, J. I., J. L. Whitmore, H. E. Wahlgren, J. F. Laundrie, and K. A. Christopherson. 1980. *Forestry Activities and Deforestation Problems in Developing Countries* (Washington, Forest Products Laboratory, USDA Forest Service).

Appendixes

Appendixes

APPENDIXES

Appendix A

REPRESENTATIVE PLANTATION REGIMES: COSTS AND YIELDS

This appendix presents the representative pulpwood and integrated (sawtimber) regimes (see chapter 2) that were used as the input for the Financial Return Program for Forestry Investments (Goforth and Mills, 1975). The stumpage prices given are those used in the base case. The stumpage prices that were used in the alternative scenarios for the sensitivity analyses are given in appendix D.

The integrated regimes with lower-quality sawtimber (limited to softwood regimes in nontraditional producing regions) used the same input as the associated integrated regime, but with lower sawtimber shadow stumpage prices. The sawtimber market prices were reduced 10 percent in the calculation of these stumpage prices. These reduced sawtimber stumpage prices are given in the same table as the associated integrated regime.

Yields shown are merchantable solid wood yields, inside bark. In general, yields do not reflect the gains possible through intensive genetic improvement of planting stock (Dutrow and Row, 1976).

All yields and costs presented are on a per hectare basis. Costs are in 1979 U.S. dollars, and stumpage prices are in 1979 U.S. dollars per cubic meter.

TABLE OF CONTENTS, APPENDIX A

Region/species	Pulpwood	Regime Integrated, with standard-quality sawtimber	Integrated, with lower-quality sawtimber
	Table	Table	Table
North America			
U.S. South			
Pinus taeda, avg.-yield site	A-1	A-2	--
Pinus taeda, high-yield site	A-3	A-4	--
Pacific Northwest			
Pseudotsuga menziesii, avg.-yield site	A-5	A-6	--
Pseudotsuga menziesii, high-yield site	A-7	A-8	--
South America			
Brazil, Amazonia			
Pinus caribaea	A-9	A-10	A-10
Gmelina spp.	A-11	A-12	--
Brazil, Central			
Eucalyptus spp.	A-13	A-14	--
Brazil, Southern			
Pinus taeda	A-15	A-16	A-16
Chile			
Pinus radiata	A-17	A-18	A-18
Oceania			
Australia			
Pinus radiata	A-19	A-20	A-20
New Zealand			
Pinus radiata	A-21	A-22	A-22
Africa			
South Africa			
Pinus patula	A-23	A-24	A-24
Gambia-Senegal			
Gmelina spp.	A-25	A-26	--
Eucalyptus spp.	A-27	A-28	--
Europe			
Nordic			
Picea abies	A-29	A-30	--
Asia			
Borneo			
Pinus caribaea	A-31	A-32	A-32

Table A-1. North America, U.S. South, <u>Pinus taeda</u>, Average-Yield-Site Pulpwood Regime: Management Practices, with Associated Costs, Stumpage Prices, and Yields

Year	Practice	Cost (per hectare)	Stumpage price (per m^3)	Yield (per hectare)
0	Site preparation	$180.75	--	--
0	Planting	83.42	--	--
0-30	Stand protection	2.00 ea. yr.	--	--
4	Hardwood control	69.52	--	--
10	Controlled burn	11.63	--	--
17	Pulpwood commercial thin	--	$14.94	49.31 m^3
22	Pulpwood commercial thin	--	14.94	49.31 m^3
30	Pulpwood harvest	--	18.94	258.05 m^3

Sources and Notes:

Site preparation, planting, and hardwood control costs from Dutrow (1978). Hardwood control costs are cited as "cleaning and release." 1978 costs for Southeast updated to 1979 using 1978-79 WPIs, International Monetary Fund (1980, line 63). Stand protection costs estimated from Bellinger (1981) and Weyerhaeuser (1981). Prescribed burn costs average of Weyerhaeuser (1979) and Mills and Cain (1978).

Regime from Coile and Schumacher (1964).

Yields are from Coile and Schumacher (1964)--Coastal Plain, old-field, site indices 60 and 70 at age 25 for medium and high sites, respectively, 800 trees per acre at age 5. A conversion factor of 78.3 cubic feet per cord was used, which produced yields that are an approximate average of those published in Burkhart and coauthors (1972), Coile and Schumacher (1964), Feduccia and coauthors (1979), Smalley and Bailey (1974), and Smith (1976) (with unpublished yield tables).

Table A-2. North America, U.S. South, <u>Pinus taeda</u>, Average-Yield-Site Integrated Regime: Management Practices, with Associated Costs, Stumpage Prices, and Yields

Year	Practice	Cost (per hectare)	Stumpage price (per m^3)	Yield (per hectare)
0	Site preparation	$180.75	--	--
0	Planting	83.42	--	--
0-35	Stand protection	2.00 ea. yr.	--	--
4	Hardwood control	69.52	--	--
10	Controlled burn	11.63	--	--
17	Pulpwood commercial thin	--	14.94	49.31 m^3
22	Pulpwood commercial thin	--	14.94	49.31 m^3
30	Pulpwood commercial thin	--	14.94	33.43 m^3
30	Sawtimber commercial thin	--	27.88	15.89 m^3
35	Pulpwood harvest	--	18.94	23.73 m^3
35	Sawtimber harvest	--	31.88	262.26 m^3

<u>Sources and Notes:</u>

Site preparation, planting and hardwood control costs from Dutrow (1978). Hardwood control costs are cited as "cleaning and release." 1978 costs for Southeast updated to 1979 using 1978-79 WPIs, International Monetary Fund (1980, line 63). Stand protection costs estimated from Bellinger (1981) and Weyerhaeuser (1981). Prescribed burn costs average of Weyerhaeuser (1979) and Mills and Cain (1978).

Regime from Coile and Schumacher (1964).

Yields are from Coile and Schumacher (1964)--Coastal Plain, oldfield, site indices 60 and 70 at age 25 for medium and high sites, respectively, 800 trees per acre at age 5. A conversion factor of 78.3 cubic feet per cord was used, which produced yields that are an approximate average of those published in Burkhart and coauthors (1972), Coile and Schumacher (1964), Feduccia and coauthors (1979), Smalley and Bailey (1974), and Smith (1976) (with unpublished yield tables). Yields allocated to sawtimber and pulpwood yield roughly in the same proportion as that found in Mills and Cain (1978).

Table A-3. North America, U.S. South, Pinus taeda, High-Yield-Site Pulpwood Regime: Management Practices, with Associated Costs, Stumpage Prices and Yields

Year	Practice	Cost (per hectare)	Stumpage price (per m^3)	Yield (per hectare)
0	Site preparation	$180.75	--	--
0	Planting	83.42	--	--
0-30	Stand protection	2.00 ea. yr.	--	--
4	Hardwood control	69.52	--	--
10	Controlled burn	11.63	--	--
15	Pulpwood commercial thin	--	14.94	35.35 m^3
20	Pulpwood commercial thin	--	14.94	54.79 m^3
25	Pulpwood commercial thin	--	14.94	71.23 m^3
30	Pulpwood harvest	--	18.94	373.65 m^3

Sources and Notes:

Site preparation, planting, and hardwood control costs from Dutrow (1978). Hardwood control costs are cited as "cleaning and release." 1978 "medium" costs for Southeast updated to 1979 using 1978-79 WPIs, International Monetary Fund (1980, line 63). Stand protection costs estimated from Bellinger (1981) and Weyerhaeuser (1981). Prescribed burn costs average of Weyerhaeuser (1979) and Mills and Cain (1978).

Regime from Coile and Schumacher (1964).

Yields are from Coile and Schumacher (1964)--Coastal Plain, oldfield, site indices 60 and 70 at age 25 for medium and high sites, respectively, 800 trees per acre at age 5. A conversion factor of 78.3 cubic feet per cord was used, which produced yields that are an approximate average of those published in Burkhart and coauthors (1972), Coile and Schumacher (1964), Feduccia and coauthors (1979), Smalley and Bailey (1974), and Smith (1976) (with unpublished yield tables).

Table A-4. North America, U.S. South, Pinus taeda, High-Yield-Site Integrated Regime: Management Practices, with Associated Costs, Stumpage Prices, and Yields

Year	Practice	Cost (per hectare)	Stumpage price (per m^3)	Yield (per hectare)
0	Site preparation	$180.75	--	--
0	Planting	83.42	--	--
0-35	Stand protection	2.00 ea. yr.	--	--
4	Hardwood control	69.52	--	--
10	Controlled burn	11.63	--	--
15	Pulpwood commercial thin	--	$14.94	38.35 m^3
20	Pulpwood commercial thin	--	14.94	54.79 m^3
25	Pulpwood commercial thin	--	14.94	71.23 m^3
30	Pulpwood commercial thin	--	14.94	26.36 m^3
30	Sawtimber commercial thin	--	27.88	44.87 m^3
35	Pulpwood harvest	--	18.94	31.92 m^3
35	Sawtimber harvest	--	31.88	352.68 m^3

Sources and Notes:

Site preparation, planting, and hardwood control costs from Dutrow (1978). Hardwood control costs are cited as "cleaning and release." 1978 costs for Southeast updated to 1979 using 1978-79 WPIs, International Monetary Fund (1979, line 63). Stand protection costs estimated from Bellinger (1981) and Weyerhaeuser (1981). Prescribed burn costs average of Weyerhaeuser (1979) and Mills and Cain (1978).

Regime extrapolated from Coile and Schumacher (1964).

Yields are from Coile and Schumacher (1964)--Coastal Plain, old-field, site indices 60 and 70 at age 25 for medium and high sites, respectively, 800 trees per acre at age 5. A conversion factor of 78.3 cubic feet per cord was used, which produced yields that are an approximate average of those published in Burkhart and coauthors (1972), Coile and Schumacher (1964), Feduccia and coauthors (1979), Smalley and Bailey (1974), and Smith (1976) (with unpublished yield tables). Yields allocated to sawtimber and pulpwood yield roughly in the same proportion as that found in Mills and Cain (1978).

Table A-5. North America, Pacific Northwest, Pseudotsuga menziesii, Average-Yield-Site Pulpwood Regime: Management Practices, with Associated Costs, Stumpage Prices, and Yields

Year	Practice	Cost (per hectare)	Stumpage price (per m^3)	Yield (per hectare)
0	Planting	$325.14	--	--
0-40	Stand protection	2.50 ea. yr.	--	--
1	Mountain beaver control	13.84	--	--
4	Brush control	67.19	--	--
20,30	Fertilization	108.38 ea. yr.	--	--
40	Pulpwood harvest	--	$14.26	510.00 m^3

Sources and Notes:

Regime is a composite from Larsen (1977) and Koss and Scott (1978) modified by judgment.

Yields from Curtis (unpublished DFSIM output). See Curtis and co-authors (1981).

Stand protection costs estimated from Bellinger (1981) and Weyerhaeuser (1981). Pest control costs from Weyerhaeuser (1979). Planting, brush control, and fertilization costs from Larsen and Wadsworth (1981) adjusted to 1979 using 1979-81 WPIs, International Monetary Fund (1982, line 63).

Table A-6. North America, Pacific Northwest, Pseudotsuga menziesii, Average-Yield-Site Integrated Regime: Management Practices, with Associated Costs, Stumpage Prices, and Yields

Year	Practice	Cost (per hectare)	Stumpage price (per m^3)	Yield (per hectare)
0	Planting	$325.14	--	--
0-50	Stand protection	2.50 ea. yr.	--	--
1	Mountain beaver control	13.84	--	--
4	Brush control	67.19	--	--
15	Precommercial thin	184.25	--	--
20,30,40	Fertilization	108.38 ea. yr.	--	--
50	Pulpwood harvest	--	$14.26	73.00 m^3
50	Sawtimber harvest	--	23.79	648.00 m^3

Sources and Notes:

Regime is a composite from Larsen (1977) and Koss and Scott (1978) modified by judgment.

Yields from Curtis (unpublished DFSIM output). See Curtis and co-authors (1981).

Stand protection costs estimated from Bellinger (1981) and Weyerhaeuser (1981). Pest control costs from Weyerhaeuser (1979). Planting, brush control, precommercial thin, and fertilization costs from Larsen and Wadsworth (1981) adjusted to 1979 using 1979-81 WPIs, International Monetary Fund (1982, line 63).

Table A-7. North America, Pacific Northwest, Pseudotsuga menziesii, High-Yield-Site Pulpwood Regime: Management Practices, with Associated Costs, Stumpage Prices, and Yields

Year	Practice	Cost (per hectare)	Stumpage price (per m^3)	Yield (per hectare)
0	Planting	$325.14	--	--
0-30	Stand protection	2.50 ea. yr.	--	--
1	Mountain beaver control	13.84	--	--
4	Brush control	67.19	--	--
15,25	Fertilization	108.38 ea. yr.	--	--
30	Pulpwood harvest	--	$14.26	408.00 m^3

Sources and Notes:

Regime is a composite from Larsen (1977) and Koss and Scott (1978) modified by judgment.

Yields from Curtis (unpublished DFSIM output). See Curtis and co-authors (1981).

Stand protection costs estimated from Bellinger (1981) and Weyerhaeuser (1981). Pest control costs from Weyerhaeuser (1979). Planting, brush control, and fertilization costs from Larsen and Wadsworth (1981) adjusted to 1979 using 1979-81 WPIs, International Monetary Fund (1982, line 63).

Table A-8. North America, Pacific Northwest, Pseudotsuga menziesii, High-Yield-Site Integrated Regime: Management Practices, with Associated Costs, Stumpage Prices, and Yields

Year	Practice	Cost (per hectare)	Stumpage price (per m^3)	Yield (per hectare)
0	Planting	$325.14	--	--
0-40	Stand protection	2.50 ea. yr.	--	--
1	Mountain beaver control	13.84	--	--
4	Brush Control	67.19	--	--
15	Precommercial thin	184.25	--	--
15,25,35	Fertilization	108.38 ea. yr.	--	--
30	Pulpwood commercial thin	--	$10.26	110.00 m^3
30	Sawtimber commercial thin	--	19.79	44.00 m^3
40	Pulpwood harvest	--	14.26	66.00 m^3
40	Sawtimber harvest	--	23.79	581.00 m^3

Sources and Notes:

Regime is a composite from Larsen (1977) and Koss and Scott (1978) modified by judgment.

Yields from Curtis (unpublished DFSIM output). See Curtis and co-authors (1981).

Stand protection costs estimated from Bellinger (1981) and Weyerhaeuser (1981). Pest control costs from Weyerhaeuser (1979). Planting, brush control, precommercial thin, and fertilization costs from Larsen and Wadsworth (1981) adjusted to 1979 using 1979-81 WPIs, International Monetary Fund (1982, line 63).

Table A-9. South America, Brazil, Amazonia, Pinus caribaea, Pulpwood Regime: Management Practices, with Associated Costs, Stumpage Prices, and Yields

Year	Practice	Cost (per hectare)	Stumpage price (per m^3)	Yield (per hectare)
0	Site preparation	$55.00	--	--
0	Controlled burn	7.50	--	--
0	Planting stock	160.00	--	--
0	Planting	40.00	--	--
0-12	Stand protection	15.00 ea. yr.	--	--
1	Replanting seedling loss	5.00	--	--
1-4	Clean and release	50.00 ea. yr.	--	--
12	Pulpwood harvest	--	$18.26	192.00 m^3

Source and Notes:

Regime and costs from Jari Project (1979).

Amazon projects are not eligible for Brazilian fiscal incentives program. Thus, establishment costs used are actual Jari costs.

Table A-10. South America, Brazil, Amazonia, Pinus caribaea, Integrated Regime: Management Practices, with Associated Costs, Stumpage Prices, and Yields

Year	Practice	Cost (per hectare)	Stumpage price (per m^3)	Yield (per hectare)
0	Site preparation	$55.00	--	--
0	Controlled burn	7.50	--	--
0	Planting stock	160.00	--	--
0	Planting	40.00	--	--
0-16	Stand protection	15.00 ea. yr.	--	--
1	Replanting seedling loss	5.00	--	--
1-4	Clean and release	50.00 ea. yr.	--	--
9	Pulpwood commercial thin	--	$14.26	70.00 m^3
12	Pulpwood commercial thin	--	14.26	42.00 m^3
12	Sawtimber commercial	--	26.49	18.00 m^3
16	Pulpwood harvest	--	18.26	31.50 m^3
16	Sawtimber harvest	--	30.49	94.50 m^3

Lower-quality sawtimber stumpage prices - Commercial thin: $19.34
Harvest: 23.34

Source and Notes:

Regime and costs from Jari Project (1979).

Amazon projects are not eligible for Brazilian fiscal incentives program. Thus, establishment costs used are actual Jari costs.

Table A-11. South America, Brazil, Amazonia, <u>Gmelina</u> spp., Pulpwood Regime: Management Practices, with Associated Costs, Stumpage Prices, and Yields

Year	Practice	Cost (per hectare)	Stumpage price (per m^3)	Yield (per hectare)
0	Site preparation	$55.00	--	--
0	Controlled burn	7.50	--	--
0	Planting stock	27.50	--	--
0	Planting	40.00	--	--
0-19	Stand protection	15.00 ea. yr.	--	--
1	Replanting seedling loss	10.00	--	--
1-3	Clean and release	50.00 ea. yr.	--	--
7	Pulpwood harvest	--	$11.71	126.00 m^3
7	Site preparation	55.00	--	--
8,9	Clean and release	50.00 ea. yr.		
8	Unwanted sprout removal	35.00		
13	Pulpwood harvest	--	11.71	108.00 m^3
13	Site preparation	55.00	--	--
14,15	Clean and release	50.00 ea. yr.	--	--
14	Unwanted sprout removal	35.00	--	--
19	Pulpwood harvest	--	11.71	108.00 m^3

Source and Notes:

Regime and costs from Jari Project (1979).

Amazon projects are not eligible for Brazilian fiscal incentives program. Thus, establishment costs used are actual Jari costs.

Table A-12. South America, Brazil, Amazonia, Gmelina spp., Integrated Regime: Management Practices, with Associated Costs, Stumpage Prices, and Yields

Year	Practice	Cost (per hectare)	Stumpage price (per m^3)	Yield (per hectare)
0	Site preparation	$55.00	--	--
0	Controlled burn	7.50	--	--
0	Planting stock	27.50	--	--
0	Planting	40.00	--	--
0-12	Stand protection	15.00 ea. yr.	--	--
1	Replanting seedling loss	10.00	--	--
1	Clean and release	50.00	--	--
2	Mark and prune	60.00	--	--
3	Precommercial thin	30.00	--	--
4	Pruning	60.00	--	--
7	Pulpwood commercial thin	--	7.71	74.00 m^3
12	Pulpwood harvest	--	11.71	35.50 m^3
12	Sawtimber harvest	--	20.79	106.50 m^3

Source and Notes:

Regime and costs from Jari Project (1979).

Amazon projects are not eligible for Brazilian fiscal incentives program. Thus, establishment costs used are actual Jari costs.

Table A-13. South America, Brazil, Central, Eucalyptus spp., Pulpwood Regime: Management Practices, with Associated Costs, Stumpage Prices, and Yields

Year	Practice	Cost (per hectare)	Stumpage price (per m^3)	Yield (per hectare)
0	Stand establishment	$508.00	--	--
0-19	Stand protection	15.00 ea. yr.	--	--
1	Clean and release	109.00	--	--
2	Clean and release	72.50	--	--
3	Clean and release	36.30	--	--
7	Pulpwood harvest	--	$11.71	175.00 m^3
7	Sprout removal	10.00	--	--
8	Clean and release	70.00	--	--
13	Pulpwood harvest	--	11.71	150.00 m^3
13	Sprout removal	10.00	--	--
14	Clean and release	70.00	--	--
19	Pulpwood harvest	--	11.71	150.00 m^3

Source and Notes:

Regime and costs from Associação Nacional dos Fabricantes de Papel e Celulose (1979).

Establishment costs are at the level established by the Brazilian fiscal incentives program.

Table A-14. South America, Brazil, Central, Eucalyptus spp., Integrated Regime: Management Practices, with Associated Costs, Stumpage Prices, and Yields

Year	Practice	Cost (per hectare)	Stumpage price (per m^3)	Yield (per hectare)
0	Stand establishment	$508.00	--	--
0-19	Stand protection	15.00 ea. yr.	--	--
1	Clean and release	109.00	--	--
2	Clean and release	72.50	--	--
3	Clean and release	36.30	--	--
7	Pulpwood commercial thin	--	$7.71	85.00 m^3
7	Sprout removal	10.00	--	--
8	Clean and release	70.00	--	--
13	Pulpwood commercial thin	--	7.71	45.00 m^3
13	Sawtimber commercial thin	--	16.79	70.00 m^3
13	Sprout removal	10.00	--	--
14	Clean and relrease	70.00	--	--
19	Pulpwood harvest	--	11.71	30.00 m^3
19	Sawtimber harvest	--	20.79	245.00 m^3

Source and Notes:

Regime and costs from Associacão Nacionál dos Fabricantes de Papel e Celulose (1979).

Establishment costs are at the level established by the Brazilian fiscal incentives program.

Table A-15. South America, Brazil, Southern, <u>Pinus taeda</u>, Pulpwood Regime: Management Practices, with Associated Costs, Stumpage Prices, and Yields

Year	Practice	Cost (per hectare)	Stumpage price (per m^3)	Yield (per hectare)
0	Stand establishment	$508.00	--	--
0-12	Stand protection	15.00 ea. yr.	--	--
1	Clean and release	109.00	--	--
2	Clean and release	72.50	--	--
3	Clean and release	36.30	--	--
12	Pulpwood harvest	--	$18.26	240.00 m^3

Sources and Notes:

Regime and costs from Associaçao Nacionál dos Fabricantes de Papel e Celulose (1979).

Establishment costs are at the level established by the Brazilian fiscal incentives program.

Pulpwood rotation age used is same as Brazil, Amazon, <u>P. caribaea</u>, pulpwood regime.

Table A-16. South America, Brazil, Southern, Pinus taeda, Integrated Regime: Management Practices, with Associated Costs, Stumpage Prices, and Yields

Year	Practice	Cost (per hectare)	Stumpage price (per m^3)	Yield (per hectare)
0	Stand establishment	$508.00	--	--
0-20	Stand protection	15.00 ea. yr.	--	--
1	Clean and release	109.00	--	--
2	Clean and release	72.50	--	--
3	Clean and release	36.30	--	--
8	Pulpwood commercial thin	--	$14.26	80.00 m^3
12	Pulpwood commercial thin	--	14.26	80.00 m^3
16	Pulpwood commercial thin	--	14.26	56.00 m^3
16	Sawtimber commercial thin	--	26.49	24.00 m^3
20	Pulpwood harvest	--	18.26	13.00 m^3
20	Sawtimber harvest	--	30.49	147.00 m^3

Lower-quality sawtimber stumpage prices - Commercial thin: $19.34
Harvest: 23.34

Sources and Notes:

Regime and costs from Associaçao Nacional dos Fabricantes de Papel e Celulose (1979).

Establishment costs are at the level established by the Brazilian fiscal incentives program.

Table A-17. South America, Chile, Pinus radiata, Pulpwood Regime: Management Practices, with Associated Costs, Stumpage Prices, and Yields

Year	Practice	Cost (per hectare)	Stumpage price (per m^3)	Yield (per hectare)
0	Site preparation and stand establishment	$173.49	--	--
0-25	Stand protection	3.13 ea. yr.	--	--
1	Replanting seedling loss	17.35	--	--
10	Pulpwood commercial thin	--	11.45	121.00 m^3
25	Pulpwood harvest	--	15.45	429.00 m^3

Sources:

Regime and costs from Schlatter (1976) and Corporación Nacional Forestal (1979)

Table A-18. South America, Chile, Pinus radiata, Integrated Regime: Management Practices, with Associated Costs, Stumpage Prices, and Yields

Year	Practice	Cost (per hectare)	Stumpage price (per m^3)	Yield (per hectare)
0	Site preparation and stand establishment	$173.49	--	--
0-32	Stand protection	3.13 ea. yr.	--	--
1	Replanting seedling loss	17.35	--	--
18	Pulpwood commercial thin	--	11.45	52.80 m^3
18	Sawtimber commercial thin	--	20.75	54.20 m^3
25	Pulpwood commercial thin	--	11.45	39.40 m^3
25	Sawtimber commercial thin	--	20.75	128.10 m^3
32	Pulpwood harvest	--	15.45	30.30 m^3
32	Sawtimber harvest	--	24.75	398.50 m^3

Lower-quality sawtimber stumpage prices - Commercial thin: $13.60
Harvest: 17.60

Sources:

Regime and costs from Schlatter (1976) and Corporación Nacional Forestal (1979).

Table A-19. Oceania, Australia, <u>Pinus radiata</u>, Pulpwood Regime: Management Practices, with Associated Costs, Stumpage Prices, and Yields

Year	Practice	Cost (per hectare)	Stumpage price (per m^3)	Yield (per hectare)
0	Fell and heap (clear-cut and windrow)	$313.80	--	--
0	Controlled burn	11.11	--	--
0	Plow	52.76	--	--
0	Planting stock	55.54	--	--
0	Planting	122.19	--	--
0	Fertilization	88.86	--	--
0	Pest control	4.17	--	--
0-29	Stand protection	3.13 ea. yr	--	--
1	Replanting seedling loss	4.17	--	--
2	Clean and release	81.92	--	--
11	Pulpwood commercial thin	--	$9.66	130.00 m^3
29	Pulpwood harvest	--	13.66	470.00 m^3

Sources and Notes:

Establishment regime and costs from Greig (1979a and 1979b). Thinning and harvest ages from Chavasse (1977). Yields from Unwin (1979). ("Scantling" regime on site 28).

Pulpwood stumpage prices are average of all regions' reported royalties for pulpwood. Costs and prices expanded from 1977 Australian $ to 1979 Australian $ using Australian WPI, International Monetary Fund (1980, line 63, p. 91) and exchanged to 1979 U.S. $ using 1979 Australian to U.S. exchange rate, International Monetary Fund (1980, line rf, p. 91).

Table A-20. Oceania, Australia, Pinus radiata, Integrated Regime: Management Practices, with Associated Costs, Stumpage Prices, and Yields

Year	Practice	Cost (per hectare)	Stumpage price (per m^3)	Yield (per hectare)
0	Fell and heap (clear-cut and windrow)	$313.80	--	--
0	Controlled burn	11.11	--	--
0	Plow	52.76	--	--
0	Planting stock	55.54	--	--
0	Planting	122.19	--	--
0	Fertilization	88.86	--	--
0	Pest control	4.17	--	--
0-35	Stand protection	3.13 ea. yr.	--	--
1	Replanting seedling loss	4.17	--	--
2	Clean and release	81.92	--	--
13	Pulpwood commercial thin	--	$9.66	77.00 m^3
20	Pulpwood commercial thin	--	9.66	40.00 m^3
20	Sawtimber commercial thin	--	17.09	41.00 m^3
27	Pulpwood commercial thin	--	9.66	30.00 m^3
27	Sawtimber commercial thin	--	17.09	97.00 m^3
35	Pulpwood harvest	--	13.66	23.00 m^3
35	Sawtimber harvest	--	21.09	301.00 m^3

Lower-quality sawtimber stumpage prices - Commercial thin: $10.58
Harvest: 14.58

Sources and Notes:

Regime, costs, and prices from Greig (1979a and 1979b). All yields from trees of DBH of 20 cm or less are considered pulpwood (Greig, 1979b).

Pulpwood stumpage prices are average of all regions' reported royalties for pulpwood. Sawtimber stumpage price is average of royalties reported for all size classes greater than 20 cm (excluding peelers), for all regions (Greig, 1979b). Costs and prices expanded from 1977 Australian $ to 1979 Australian $ using Australian WPI, International Monetary Fund (1980, line 63, p. 91) and exchanged to 1979 U.S. $ using 1979 Australian to U.S. exchange rate, International Monetary Fund (1980, line rh, p. 91).

Table A-21. Oceania, New Zealand, *Pinus radiata*, Pulpwood Regime: Management practices, with Associated Costs, Stumpage Prices, and Yields

Year	Practice	Cost (per hectare)	Stumpage price (per m^3)	Yield (per hectare)
0	Site preparation	$250.04	--	--
0	Planting stock	84.39	--	--
0	Planting	118.77	--	--
0-18	Stand protection	3.13 ea. yr.	--	--
1	Replanting seedling loss	39.07	--	--
1,2	Clean and release	176.59 ea. yr.	--	--
18	Pulpwood harvest	--	13.66	450.00 m^3

Sources and Notes:

Regime and yields from Fenton and Tennent (1976).

Costs from Walker (1977). Costs updated from 1976 NZ $ to 1979 NZ $ using NZ WPI, International Monetary Fund (1980, line 63, p. 309) and exchanged to 1979 U.S. $ using 1979 NZ to U.S. exchange rate, International Monetary Fund (1980, line rh, p. 309).

Table A-22. Oceania, New Zealand, Pinus radiata, Integrated Regime: Management Practices, with Associated Costs, Stumpage Prices, and Yields

Year	Practice	Cost (per hectare)	Stumpage price (per m^3)	Yield (per hectare)
0	Site preparation	$250.04	--	--
0	Planting stock	84.39	--	--
0	Planting	118.77	--	--
0-27	Stand protection	3.13. ea. yr.	--	--
1	Replanting seedling loss	39.07	--	--
1,2	Clean and release	176.59 ea. yr	--	--
8	Precommercial thin	93.77	--	--
14	Pulpwood commercial thin	--	9.66	76.50 m^3
14	Sawtimber commercial thin	--	17.09	73.50 m^3
20	Pulpwood commercial thin	--	9.66	42.00 m^3
20	Sawtimber commercial thin	--	17.09	133.00 m^3
27	Pulpwood harvest	--	13.66	24.50 m^3
27	Sawtimber harvest	--	21.09	325.50 m^3

Lower-quality sawtimber stumpage prices - Commercial thin: $10.58
Harvest: 14.58

Sources and Notes:

Regime and yields from Fenton and Tennent (1976). Pulpwood/sawtimber proportions based on Australian integrated regime.

Costs from Walker (1977). Costs updated from 1976 NZ $ to 1979 NZ $ using NZ WPI, International Monetary Fund (1980, line 63, p. 309) and exchanged to 1979 U.S. $ using 1979 NZ to U.S. exchange rate, International Monetary Fund (1980, line rh, p. 309).

Table A-23. Africa, South Africa, Pinus patula, Pulpwood Regime: Management Practices, with Associated Costs, Stumpage Prices, and Yields

Year	Practice	Cost (per hectare)	Stumpage price (per m^3)	Yield (per hectare)
0	Site preparation	$89.37	--	--
0	Planting stock	51.26	--	--
0	Planting	37.06	--	--
0-15	Stand protection	7.00 ea. yr.	--	--
1	Replanting seedling loss	19.25	--	--
1,2	Weeding	26.30 ea. yr.	--	--
7	Pruning	25.15	--	--
15	Pulpwood harvest	--	16.55	241.50 m^3

Sources and Notes:

Regime and yields generally from Streyffert (1968), and Wormald (1975).

Costs from Grut (1970).

1965/66 Rands updated to 1979 Rands using South African WPI, International Monetary Fund (1980, line 63, p. 375) and exchanged to 1979 U.S. $ using 1979 Rand to U.S. $ exchange rate, International Monetary Fund (1980, line rh, p. 373).

Table A-24. Africa, South Africa, Pinus patula, Integrated Regime: Management Practices, with Associated Costs, Stumpage Prices, and Yields

Year	Practice	Cost (per hectare)	Stumpage price (per m^3)	Yield (per hectare)
0	Site preparation	$89.37	--	--
0	Planting stock	51.26	--	--
0	Planting	37.06	--	--
0-26	Stand protection	7.00 ea. yr.	--	--
1	Replanting seedling loss	19.25	--	--
1,2	Weeding	26.30 ea. yr.	--	--
6,8	Pruning	25.15 ea. yr.	--	--
6	Pulpwood commercial thin	--	12.55	9.30 m^3
10	Pruning	21.15	--	--
11	Pulpwood commercial thin	--	12.55	15.60 m^3
11	Sawtimber commercial thin	--	23.01	3.90 m^3
16	Pulpwood commercial thin	--	12.55	27.80 m^3
16	Sawtimber commercial thin	--	23.01	27.80 m^3
21	Pulpwood commercial thin	--	12.55	16.20 m^3
21	Sawtimber commercial thin	--	23.01	37.90 m^3
26	Pulpwood harvest	--	16.55	41.60 m^3
26	Sawtimber harvest	--	27.01	235.90 m^3

Lower-quality sawtimber stumpage prices - Commercial thin: $15.86
Harvest: 19.86

Sources and Notes:

Regime and yields from Grut (1970; see p. 194 for yields.) Pulpwood/sawtimber proportions roughly from Wormald (1975, p. 145).

Costs from Grut (1970).

1965/66 Rands updated to 1979 Rands using South African WPI, International Monetary Fund (1980, line 63, p. 375) and exchanged to 1979 U.S. $ using 1979 Rand to U.S. $ exchange rate, International Monetary Fund (1980, line rh, p. 373).

Table A-25. Africa, Gambia-Senegal, Gmelina spp., Pulpwood Regime: Management Practices, with Associated Costs, Stumpage Prices, and Yields

Year	Practice	Cost (per hectare)	Stumpage price (per m^3)	Yield (per hectare)
0	Stand establishment	$345.00	--	--
0-30	Stand protection	5.00 ea. yr.	--	--
1	Weeding	35.00	--	--
10	Pulpwood harvest	--	$12.29	150.00 m^3
10	Weeding	35.00	--	--
20	Pulpwood harvest	--	12.29	150.00 m^3
20	Weeding	35.00	--	--
30	Pulpwood harvest	--	12.29	150.00 m^3

Source: Regime, yields, and costs from Robert R. Nathan Associates (1979).

Table A-26. Africa, Gambia-Senegal, Gmelina spp., Integrated Regime: Management Practices, with Associated Costs, Stumpage Prices, and Yields

Year	Practice	Cost (per hectare)	Stumpage price (per m^3)	Yield (per hectare)
0	Stand establishment	$345.00	--	--
0-40	Stand protection	5.00 ea. yr.	--	--
1	Weeding	35.00	--	--
5	Pulpwood commercial thin	--	$8.29	30.00 m^3
10	Pulpwood commercial thin	--	8.29	54.00 m^3
15	Pulpwood commercial thin	--	8.29	35.00 m^3
15	Sawtimber commercial thin	--	17.87	35.00 m^3
20	Pulpwood harvest	--	12.29	36.00 m^3
20	Sawtimber harvest	--	21.87	110.00 m^3
20	Weeding	35.00	--	--
25	Pulpwood commercial thin	--	8.29	30.00 m^3
30	Pulpwood commercial thin	--	8.29	54.00 m^3
35	Pulpwood commercial thin	--	8.29	35.00 m^3
35	Sawtimber commercial thin	--	17.87	35.00 m^3
40	Pulpwood harvest	--	12.29	36.00 m^3
40	Sawtimber harvest	--	21.87	110.00 m^3

Source: Regime, yields, and costs from Robert R. Nathan Associates (1979).

Table A-27. Africa, Gambia-Senegal, Eucalyptus spp., Pulpwood Regime: Management Practices, with Associated Costs, Stumpage Prices, and Yields

Year	Practice	Cost (per hectare)	Stumpage price (per m^3)	Yield (per hectare)
0	Stand establishment	$956.00	--	--
0-21	Stand protection	5.00 ea. yr.	--	--
1,2	Weeding	35.00 ea. yr.	--	--
7	Pulpwood harvest	--	$12.29	119.00 m^3
7	Weeding	35.00	--	--
14	Pulpwood harvest		12.29	119.00 m^3
14	Weeding	35.00	--	--
21	Pulpwood harvest	--	12.29	119.00 m^3

Source: Regime, yields, and costs from Robert R. Nathan Associates (1979).

Table A-28. Africa, Gambia-Senegal, Eucalyptus spp., Integrated Regime: Management Practices, with Associated Costs, Stumpage Prices, and Yields

Year	Practice	Cost (per hectare)	Stumpage price (per m^3)	Yield (per hectare)
0	Stand establishment	$956.00	--	--
0-30	Stand protection	5.00 ea. yr.	--	--
1,2	Weeding	35.00 ea. yr.	--	--
5	Pulpwood commercial thin	--	8.29	34.00 m^3
10	Pulpwood harvest	--	12.29	34.00 m^3
10	Sawtimber harvest	--	21.87	102.00 m^3
10,11	Weeding	35.00 ea. yr.	--	--
15	Pulpwood commercial thin	--	8.29	34.00 m^3
20	Pulpwood harvest	--	12.29	34.00 m^3
20	Sawtimber harvest	--	21.87	102.00 m^3
20,21	Weeding	35.00 ea. yr.	--	--
25	Pulpwood commercial thin	--	8.29	34.00 m^3
30	Pulpwood harvest	--	12.29	34.00 m^3
30	Sawtimber harvest	--	21.87	102.00 m^3

Source: Regime, yields, and costs from Robert R. Nathan Associates (1979).

Table A-29. Europe, Nordic, **Picea abies**, Pulpwood Regime: Management Practices, with Associated Costs, Stumpage Prices, and Yields

Year	Practice	Cost (per hectare)	Stumpage price (per m^3)	Yield (per hectare)
0	Site preparation	$183.50	--	--
0	Planting	225.60	--	--
0	Fertilization	46.54	--	--
0-50	Stand protection	1.00 ea. yr.	--	--
1	Replant seedling loss	77.72	--	--
50	Pulpwood harvest	--	$20.98	250.00 m^3

Sources and Notes:

Regime, costs, and yields from Finland (1977, pp. 98, 136). Pulpwood regime derived from modification of integrated regime.

Costs updated from 1975 Finnish marks to 1979 U.S. $ using 1975-79 Finnish Price Index and 1979 US/Finland exchange rate, International Monetary Fund (1979).

Table A-30. Europe, Nordic, Picea abies, Integrated Regime: Management Practice, with Associated Costs, Stumpage Prices, and Yields

Year	Practice	Cost (per hectare)	Stumpage price (per m^3)	Yield (per hectare)
0	Site preparation	$183.50	--	--
0	Planting	225.60	--	--
0	Fertilization	46.54	--	--
0-80	Stand protection	1.00 ea. yr.		
1	Replanting seedling loss	77.72	--	--
25	Pulpwood commercial thin	--	$16.98	60.00 m^3
50	Pulpwood commercial thin	--	16.98	22.50 m^3
50	Sawtimber commercial thin	--	32.06	67.50 m^3
80	Pulpwood harvest	--	20.98	12.50 m^3
80	Sawtimber harvest	--	36.06	237.50 m^3

Sources and Notes:

Regime, costs, and yields from Finland (1977, pp. 98, 136).

Sawtimber/pulpwood mix based on:

1. Thinning--25 percent pulpwood, 75 percent sawtimber
2. Clearcut--5 percent pulpwood, 95 percent sawtimber

Costs updated from 1975 Finnish marks to 1979 US $ using 1975-79 Finnish Price Index and 1979 US/Finland exchange rate, International Monetary Fund (1979).

Table A-31. Asia, Borneo, Pinus caribaea, Pulpwood Regime: Management Practices, with Associated Costs, Stumpage Prices, and Yields

Year	Practice	Cost (per hectare)	Stumpage price (per m^3)	Yield (per hectare)
0	Site preparation	$55.00	--	--
0	Controlled burn	7.50	--	--
0	Planting stock	160.00	--	--
0	Planting	40.00	--	--
0-15	Stand protection	15.00 ea. yr.	--	--
1	Replanting seedling loss	5.00	--	--
1-4	Clean and release	50.00 ea. yr.	--	--
15	Pulpwood harvest	--	15.53	210.00 m^3

Source and Notes:

Regimes and yields based on Weyerhaeuser Company (1978).

Due to lack of data on management costs, it was assumed that those for Borneo would be similar to those of the Amazon.

Table A-32. Borneo, Pinus caribaea, Integrated Regime: Management Practice, with Associated Costs, Stumpage Prices, and Yields

Year	Practice	Cost (per hectare)	Stumpage price (per m^3)	Yield (per hectare)
0	Site preparation	$55.00	--	--
0	Controlled burn	7.50	--	--
0	Planting stock	160.00	--	--
0	Planting	40.00	--	--
0-20	Stand protection	15.00 ea. yr.	--	--
1	Replanting seedling loss	5.00	--	--
1-4	Clean and release	50.00 ea. yr.	--	--
10	Pulpwood commercial thin	--	11.53	70.00 m^3
15	Pulpwood commercial thin	--	11.53	39.00 m^3
15	Sawtimber commercial thin	--	20.92	21.00 m^3
20	Pulpwood harvest	--	15.53	35.00 m^3
20	Sawtimber harvest	--	24.92	105.00 m^3

Lower-quality sawtimber stumpage prices - Commercial thin: $14.42
Harvest: 18.42

Source and Notes:

Regimes and yields based on Weyerhaeuser Company (1978).

Due to lack of data on management costs, it was assumed that those for Borneo would be similar to those of the Amazon.

REFERENCES

Associacao Nacional dos Fabricantes de Papel e Celulose. 1979. Roger Sedjo's conversation with association officials, Sao Paulo, Brazil.

Bellinger, Melvin D. 1981. USDA Forest Service, State and Private Forestry, Cooperative Fire Protection Staff. Personal conversation with Roger Sedjo, January 5.

Burkhart, H. E., R. C. Parker, M. R. Strub, and R. G. Oderwald. 1972. Yields of Oldfield Loblolly Pine Plantations. FWS-3-72 (Blacksburg, Va., Division of Forestry and Wildlife Resources, Virginia Polytechnic Institute and State University).

Chavasse, C. G., ed. 1977. Forestry Handbook (New Zealand Institute of Foresters).

Coile, T. S., and F. X. Schumacher. 1964. Soil-Site Relations, Stand Structure, and Yields of Slash and Loblolly Pine Plantations in the Southern United States (Durham, N.C., T.S. Coile, Inc.).

Corporacion Nacional Forestal. 1979. Costos de Forestacion. Resolucion Num 596, Fija (Santiago, Chile).

Curtis, Robert O. 1981. Communication with Roger Sedjo and unpublished output of DFSIM.

_____, G. W. Clendenen, and D. J. DeMars. 1981. A New Stand Simulator for Coast Douglas-fir: DFSIM User's Guide. General Technical Report PNW-128 (Portland, Oreg., USDA Forest Service, Pacific Northwest Forest and Range Experiment Station).

Dutrow, George F. 1978. "A Study of Economic Management Opportunities to Increase Timber Supplies in the Southeast United States," FOREM vol. 1, Spring, 1978, pp. 5-9 (FOREM is a publication of the School of Forestry and Environmental Studies, Duke University, Durham, N.C.).

_____, and C. Row. 1976. Measuring Financial Gains from Genetically Improved Trees. Research Paper SO-132 (New Orleans, La., USDA Forest Service, Southern Forest and Range Experiment Station).

Feduccia, D. P., T. R. Dell, W. F. Mann, T. E. Campbell, and B. H. Polmer. 1979. Yields of Unthinned Loblolly Pine Plantations on Cutover Sites in the West Gulf Region. Research Paper SO-148 (New Orleans, La., USDA Forest Service, Southern Forest and Range Experiment Station).

Fenton, R., and R. B. Tennent. 1976. "Export Log Afforestation Profitability, 1973," New Zealand Journal of Forestry Science vol. 5, no. 3, pp. 323-346.

Finland. 1977. *Yearbook of Forest Statistics: 1975. Official Statistics of Finland,* XVII A (Helsinki), p. 8.

Goforth, Marcus H., and Thomas J. Mills. 1975. *A Financial Return Program for Forestry Investments Including Sensitivity of Results to Data Errors.* Agriculture Handbook No. 488 (Washington, USDA Forest Service, GPO).

Greig, Peter J. 1979a. Personal communication to Samuel Radcliffe, April 6, 1979.

———. 1979b. "Cash Flow Analyses of the Victorian Forests Commission's Future Softwood Plantations." Mimeo. (Victoria, Australia, Division of Forest Management, Forests Commission).

Grut, Mikael. 1970. *Pinus Radiata, Growth and Economics.* (Capetown, South Africa, A. A. Balkema).

International Monetary Fund. 1979. *International Financial Statistics* vol. XXXII, no. 10 (Washington).

———. 1980. *International Financial Statistics Yearbook 1980* vol. XXXIII (Washington).

———. 1982. *International Financial Statistics* vol. XXXIII (Washington).

Jari Project. 1979. Roger Sedjo visit to and discussions with various Jari Project company representatives.

Koss, William, and B. D. Scott. 1978. *Investment Guidelines for Douglas Fir.* DNR Report No. 36 (Olympia, Wash., Department of Natural Resources).

Larsen, David N. 1977. *Phase II. Economic Analysis. Washington Forest Productivity Study.* (Olympia, Wash., Department of Natural Resources).

———, and R. K. Wadsworth. 1981. "Timber Demand and Supply Analysis for the 1980s and Future Decades in the State of Washington." Washington Forest Productivity Study unpublished report submitted to Pacific Northwest Regional Commission (Olympia, Wash., Department of Natural Resources).

Mills, Thomas J., and D. Cain. 1978. *Timber Yield and Financial Return Performance of the 1974 Forestry Incentives Program.* Research Paper RM-204 (Fort Collins, Colo., USDA Forest Service, Rocky Mountain Forest and Range Experiment Station).

Robert R. Nathan Associates. 1979. "Gambia River Basin Development Plan: Forestry Model." Contract Study for U.S. Agency for International Development, Africa Bureau (Washington).

Schlatter, Juan E. 1976. "Principales Interrogantes en Relacion al Uso Del Suelo en la Actividad Forestal de Chile." Charlas y Conferencias, Facultad de Ingenieria Forestal, Universidad Austral de Chile. No. 4 (Valdiva, Chile), pp. 1-16.

Smalley, G. W., and R. L. Bailey. 1974. <u>Yield Tables and Stand Structure for Loblolly Pine Plantations in Tennessee, Alabama, and Georgia Highlands</u>. Research paper SO-96 (New Orleans, La., USDA Forest Service, Southern Forest and Range Experiment Station).

Smith, J. L. 1976. "Volume Yields of Site Prepared Loblolly Pine Plantations in the Lower Coastal Plain of the Carolinas, Georgia and North Florida." M.S. Thesis, University of Georgia (also an unpublished computer printout).

Streyffert, Thorsten. 1968. <u>World Pulpwood, A Study in the Competitive Position of Pulpwood in Different Forest Regions</u>. (Stockholm, Sweden, Almquist and Wicksell).

Unwin, P. T. 1979. Personal communication to Samuel Radcliffe, June 22, 1979.

Walker, K. 1977. "Financial Costs Used in the Economic Evaluation of the King Country Forestry Proposal." Internal Report No. 32 (Rotorua, New Zealand, New Zealand Forest Service, Forest Research Institute, Economics of Silviculture).

Weyerhaeuser Corporation. 1979-1981. Personal conversations of Roger Sedjo and Samuel Radcliffe with various Weyerhaeuser staff members.

_____. 1978. Personal conversation of Roger Sedjo with Weyerhaeuser Company officials in Indonesia.

_____. 1979. Personal conversations of Roger Sedjo and Samuel Radcliffe with various Weyerhaeuser staff members.

_____. 1981. Personal communication between Weyerhaeuser officials and Roger Sedjo.

Wormald, T. J. 1975. <u>Pinus patula</u>. Tropical Forestry Papers No. 7 (Oxford, England, Department of Forestry).

APPENDIXES

Appendix B

MATHEMATICAL FORMULATION OF THE MODEL

The following presents mathematically the essential elements of the comparative economics plantation model.

I. Regional Pulpwood Stumpage Price

(1) $P_{sp} = \alpha P_p - (T_p + M_p) - H$

(2) $T_p = \beta_0 + \beta_1 (D)$

where:

P_{sp} = price of pulpwood stumpage, $/m3

$P_p = \bar{P}_p$ = price of woodpulp in world market, $/m.t.

T_p = international transport costs. $/m.t.

$M_p = \bar{M}_p$ = woodpulp processing costs, $/m.t.

H = harvest and local transport costs $/m

α = conversion coefficient, m.t./m

β_0 = pulpwood parameter

β_1 = pulpwood parameter

D = distance from plantation region to market in thousand miles

H = $\bar{H}_t$ if commercial thinning. $/m^3$

$\bar{H}_t$ if final harvest cost, $/m^3$

II. Regional Sawtimber Stumpage Prices

(3) $P_{ss} = V'_{wcp} + V'_{wcs} - H$

(4) $V'_{wcp} = \alpha R V_{wcp}$

(5) $V'_{wcs} = \ell R_\ell V_{wcs}$

(6) $V_{wcp} = P_p - T_p - M_p$

(7) $V_{wcs} = P_s - T_s - M_s$

(8) $T_s = g_0 + g_1 (D)$

where:

P_{ss} = price of sawtimber stumpage, $/m^3$

V'_{wcp} = value of wood content for pulpwood, $/m^3$

V'_{wcs} = value of wood content for sawtimber, $/m^3$

R_p = percentage of sawlog that becomes pulpable residual

R_ℓ = percentage of sawlog that becomes lumber

ℓ = conversion factor, MBF/m

V_{wcp} = value of wood content for plywood, $/m.t.$

V_{wcs} = value of wood content for sawtimber, $/MBF$

$P_s = \bar{P}_s$ = price of sawnwood in world market, $/MBF$

M_s = sawnwood processing costs, $/MBF$

T_s = international transport costs, $/MBF$

g_0 = sawtimber parameter

g_1 = sawtimber parameter

III. Present Net Value Criterion

$$(10) \quad PNV = \sum_{i=1}^{\infty} \frac{(R_i - C_i)}{(1+r)^i}$$

$$(11) \quad R_i = (P_{spi})(V_{pi}) + (P_{ssi})(V_{si})$$

$$(12) \quad C_i = \overline{C}_i$$

where:

PNV = present net value per hectare

r = inflation-free discount rate

R_i = gross receipt per hectare in period i

C_i = gross management costs per hectare in period i

V_{pi} = volume of pulpwood stumpage harvested in period i, m^3/h

V_{si} = volume of sawtimber stumpage harvested in period i, m^3/h

IV. World Market Model: Final Market Price Determination for Wood Pulp

$$(13) \quad P_{spw} = \alpha P_{pe} - \alpha \left[\beta_0 + \beta_1(D_1) + M_p \right] - H$$

$$(14) \quad P_{pu} = \frac{1}{\alpha}\left[P_{spw} + H \right] + \beta_0 + \beta_1(D_2) + M_p$$

$$(15) \quad P_{pe} = \frac{1}{\alpha}\left[P_{spw} + H \right] + \beta_0 + \beta_1(D_3) + M_p$$

where:

$P_{pe} = \overline{P}_{pe}$ = woodpulp price in Europe, \$/m.t.

P_{spw} = pulpwood stumpage price, Pacific Northwest, North America,

P_{pu} = wood pulp price, eastern United States, \$/m.t.

P_{pj} = wood pulp price, Japan, \$/m.t.

APPENDIXES

D_1 = shipping distance between Pacific Northwest and Europe, thousand miles

D_2 = shipping distance between Pacific Northwest and eastern United States, thousand miles

D_3 = shipping distance between Pacific Northwest and Japan, thousand miles

V. World Market: Final Market Price Determination for Lumber

$$(16)\quad P_{\ell pw} = R_\ell \{\ell P_{\ell e} - \ell [g_0 + g_1(D_1) + M_s]\} + R_p\{\alpha P_e - \alpha[0 + 1(D_1) + M_p]\} - H$$

$$(17)\quad P_{\ell u} = \frac{1}{\ell}(P_{pw} + H) + g_0 + g_1(D_2) + M_s$$

$$(18)\quad P_{\ell j} = \frac{1}{\ell}(P_{\ell pw} + H) + g_0 + g_1(D_3) + M_s$$

where:

$P_{\ell e} = \overline{P}_{se}$ = sawtimber price in Europe, $/MBF.

$P_{\ell pw}$ = sawtimber stumpage price, Pacific Northwest, North America

$P_{\ell u}$ = lumber price, eastern United States, $/MBF

$P_{\ell j}$ = lumber price, Japan, $/MBF

Appendix C

INTERNATIONAL TRANSPORT COST METHODOLOGY

International transportation costs, and particularly international ocean freight costs, are highly variable. This reflects the industry's lack of a single, formal, coherent framework of law and public policy to control its operation (Lawrence, 1972). It also reflects the cyclical nature of the merchant shipping industry, which over the years has passed through successive periods of feast and famine. Various types of shipping rates may prevail simultaneously for the same commodity. The spot market rate may be considerably different from the long-term contract rate, reflecting both the nature of current conditions within the industry and also the nature of conditions and expectations prevailing at the time that the long-term arrangements were made. In addition, various types of restrictions are imposed both by individual countries involved and by the various shipping conferences. The freight rates may also be different between routes of similar lengths, reflecting specific conditions related to the route, such as the port facilities, expected turnaround time, and also the availability or lack of a backhaul on a particular route. Given all of these considerations, it is indeed difficult to attempt to describe such a complex set of interrelationships with a simple relationship.

Attempts have been made, however, to determine statistically the major components of ocean freight rate charges. One of the more useful examples of this research--by Lipsey and Weiss (1974)--estimates the commodity structure of ocean transportation charges. In that study, using actual shipping data, the authors found that the major determinants of freight rates are the value per ton of a commodity, its bulkiness (cubic feet per ton), the distance over which it is shipped, the prevalence of small individual shipments, and the possibility of shipping the product by tanker. The statistical study was designed to look at freight rates for a variety of commodity types, utilizing cross-sectional data, and the coefficients of the variable were estimated. Our study, by contrast, does not attempt to examine the trade flows of a variety of commodities. Instead, the focus is on two rather low-valued, bulky commodities, wood pulp and sawnwood. For each of these commodities, all of the characteristics except distance used by Lipsey and Weiss to determine the freight rate are constants. Thus, our approach is to utilize distance as the principal explanatory variable determining the transport freight rate.

A more recent study (Binkley and Horner, 1981) that examined the effect of different variables upon ocean freight rates for grain found that distance, shipment size, port efficiency, and volume of trade on the route in question were the important determinants of freight rates. Our approach assumes that the long-term volumes will be such as to create conditions for optimal shipment size, port efficiency, and volume of trade. Thus, in the long term the major determinant would be distance.

We then focused on the distance variable using the data provided by Jaakko Poyry (1977) to determine a functional difference in transportation cost for bleached kraft pulp to Rotterdam from Sweden, Finland, the southern United States, and Brazil. Our preliminary estimate was combined with other data on shipping charges on various routes (Maritime Research) to provide a degree of comparability. Both linear and nonlinear variations of the relationship were employed. The specification viewed as most representative for pulp is:

$$FR = \$16 + \$4 \text{ (per thousand miles)}$$

and the relationship chosen for sawnwood is:

$$FR = 1.6 (\$16 + \$4) \text{ (per thousand miles)}$$

where FR equals the freight rate per metric ton for wood pulp and per thousand board feet (MBF) for sawnwood.

Since a thousand board feet of sawnwood converts to about a metric ton, the freight rate for the two commodities is very similar on a per-unit-weight basis. This reflects the fact that value and bulkiness characteristics, found by Lipsey and Weiss to be important, are similar for wood pulp and sawnwood, with wood pulp being a bit more valuable per unit weight, although it is somewhat less bulky. However, the greater ease of handling wood pulp suggests a lower freight rate than for lumber.

As noted, there are numerous possible modes for transporting either wood pulp or sawnwood. In general, however, wood pulp is produced in large volumes at a centrally located pulp mill. Also, the user of the wood pulp--a paper mill--will typically require large volumes of pulp. Thus, a bulk carrier can be utilized exclusively for the transport of an entire shipload from the pulp mill to the paper mill. Sawnwood, by contrast, can be produced at sawmills that are typically of much smaller scale. In addition, there are few single final markets that could absorb a large vessel's load of lumber. Therefore, whereas pulp is commonly transported en masse to final markets--paper mills--lumber is much more likely to be transported in much smaller shipments to localized markets. Typically, this could be expected to give rise to higher transport charges. For this study, however, assume that the transportation rates are quite similar between lumber and pulp. This is likely to be the case for a large integrated-plantation sawmill/pulp mill complex where appropriate volumes of the sawnwood are "piggybacked" on the same vessel as the pulp for transportation to major markets. Such a scheme is envisioned for the Jari operation in the Amazon, and might be expected to become relatively common for large integrated operations trading large volumes of pulp and sawnwood in major world markets.

Transport Cost Estimates

Table C-1 indicates the bleached kraft pulp (long fiber) and sawnwood prices at the three world market outlets based upon the transportation costs differential generated by using the Pacific Northwest as the base surplus region. For both of the commodities, the market price is highest in Europe, lower in the northeastern United States, and lowest in Japan. The low Japanese price reflects the relative accessibility of Japan to the low-cost production of the West Coast of North America.

Utilizing the conceptual framework and functional relationship developed above, tables C-2 and C-3 show the estimated transportation costs for pulp and sawnwood between the various producing regions and the three world market outlets--Japan, northeastern United States, and Western Europe. The three specific ports used were Yokohoma, New York, and Rotterdam.[1]

---

[1] The mileage used is provided by the U.S. Naval Oceanographic Office. The West Africa distance used was from Dakar, adding 1,000 miles for the European and North American shipment, and subtracting 1,000 miles for the shipping route to Japan.

Table C-1. World Price Structure (Circa 1979)

Major markets	Bleached kraft wood pulp ($ per metric ton)	Sawnwood ($ per thousand board feet)
Northeastern United States	438.4	306.4
Western Europe	450.0	325.0
Japan	431.6	295.5

Table C-2. Transport Costs for Pulp from Selected Regions to Major World Markets

($ per metric ton)

From	To Japan	To Northeastern United States	To Western Europe
U.S. South	53.2	18.0	31.6
Pacific Northwest	33.2	40.0	51.6
S. Brazil	60.8	28.0	32.8
Chile	53.2	34.4	46.0
S. Africa	50.0	43.2	40.8
W. Africa	60.4	33.2	30.4
S.E. Australia	36.0	55.6	67.2
New Zealand	36.0	50.0	61.2
Borneo	27.2	64.8	64.8
Nordic	69.6	33.2	20.0
Amazonia	60.8	28.0	32.8

Table C-3. Transport Costs for Sawnwood Revised from Selected Regions to Major World Markets

($ per thousand board feet)

From	To Japan	To Northeastern United States	To Western Europe
U.S. South	85.1	28.0	50.6
Pacific Northwest	53.1	64.0	82.6
Brazil	97.3	44.8	52.5
Chile	85.1	55.0	73.6
South Africa	80.0	69.1	65.3
West Africa	96.6	53.1	48.6
S.E. Australia	57.6	90.0	107.5
New Zealand	57.6	80.0	97.9
Borneo	43.5	103.7	103.7
Nordic	111.4	53.1	32.0
Amazonia	97.3	44.8	52.5

Table C-4. Net Prices Received in Supply Regions for Pulp

($ per metric ton)

From	To Japan	To Northeastern United States	To Western Europe
U.S. South	378.4	420.4[a]	418.4
Pacific Northeast	398.4	398.4	398.4
Brazil	370.8	410.4	417.2[a]
Chile	378.4	404.0	404.0[a]
South Africa	381.6	395.2	409.2[a]
West Africa	371.2	405.2	419.6[a]
S.E. Australia	395.6[a]	382.8	373.8
New Zealand	395.6[a]	388.4	388.8
Borneo	404.4[a]	373.6	385.2
Nordic	362.0	405.2	430.0[a]
Amazonia	370.8	410.4	417.2[a]

[a] Indicates market with highest net price which is therefore the region's major market.

Table C-5. Net Prices Received in Supply Regions for Sawnwood ($ per thousand board-feet)

From	To Japan	To Northeastern United States	To Western Europe
U.S. South	210.4	277.6[a]	274.4
Pacific Northwest	242.4	242.4	242.4
Brazil	198.2	261.6	272.5[a]
Chile	210.4	251.4	251.4[a]
South Africa	215.4	237.3	259.7[a]
West Africa	198.9	253.3	276.4[a]
S.E. Australia	237.9[a]	216.4	217.5
New Zealand	237.9[a]	226.4	227.1
Borneo	252.0[a]	202.7	221.3
Nordic	184.1	253.3	293.0[a]
Amazonia	198.2	261.6	272.5[a]

[a] Indicates market with highest net price which is therefore the region's major market.

The range of pulp transportation costs varies from a minimum of $18 per metric ton for the southern United States shipping to New York to a maximum of $69.60 per metric ton for Nordic shipping to Japan, while sawnwood rates vary from $28.80 to $111.40 per thousand board-feet. The sawnwood rates are probably somewhat lower than commonly experienced, but they are intended to reflect prices under the idealized circumstances discussed above.

Tables C-4 and C-5 provide estimates of net prices received after transportation changes for the various regional producers in the three world market outlets. These tables indicate the market outlet that generates the highest net return for each producing region. It will be noted that in several cases the market outlet region is not the market that can be reached at the lowest transport cost, but rather is the market in which the higher price offsets the additional transport cost.

Findings

The worldwide pattern of pulp and sawnwood trade flows originating in plantations that is suggested by this method appears reasonable and consistent with experience. Not surprisingly, the plantations in Southeast Asia and Oceania find their major market in Japan, where the net prices they receive are considerably above those obtained in alternative markets. The South American producers' net price is highest in the European market despite the fact that their transport costs to the North American market are somewhat lower. This reflects the somewhat higher prices received in the European market. Southern United States producers find that their net price is greatest in the northeastern United States market. However, the differential in net price for the South between the northeastern United States and the European market is small. Finally, the Nordic countries receive the highest net price by producing for the European market, which is both easily accessible and has a high price.

A further feature of the world plantation flows is found in the plantation locations examined that provide wood products to each of the principal market outlets. The Pacific Basin countries' production flows into the Japanese market. South America, Africa, and Nordic production flows into Europe, while the northeastern United States market is fed only by the U.S. South. Of course, production from the West Coast of North America is assumed to flow into all of these markets. Although this analysis was developed to look solely at plantation-produced wood flows in selected commodities, to the extent that the relative prices and transportation costs used also apply to other forest products produced from both plantation and natural forests, the general trade patterns are likely to persist.

The net prices received vary considerably. However, the variation is considerably reduced when the discussion is limited to the "best" market for each producing region. For wood pulp with a delivered price of $450 per metric ton at Rotterdam, the net price ranges from a high of $430 for the Nordic producers selling in the European market to a low of $373 per metric ton for Australian and New Zealand producers selling in the Japanese market. However, the lowest net price for a region whose best market is Europe is $409.20 per metric ton for South Africa. The regions that have the most advantageous locations as reflected in highest prices are the Nordic countries, followed by the U.S. South, and then by Amazonia and West Africa. Those with the least advantageous location are Australia, New Zealand, Chile, and Borneo.

For sawtimber, the relative rankings of the net prices are unchanged, with the net price in the "best" market ranging from $293.0 per thousand board-feet for Nordic sawnwood in the European market to $237.9 per thousand board-feet for Australian and New Zealand sawnwood in the Japanese market.

REFERENCES

Binkley, James K., and Bruce Horner. 1981. "Major Determinants Ocean Freight Rates for Grains: An Econometric Analysis," *American Journal of Agricultural Economics* vol. 63, no. 1 (February) pp. 47-57.

Jaakko Poyry. 1977. "Jaako Poyry Report: 1975-1977" (Helsinki, Finland, Jaakko Poyry Consulting Oy) pp. 5-6.

Lawrence, Samuel A. 1972. *International Sea Transport: The Years Ahead* (Lexington Books, D.C. Heath and Company).

Lipsey, Robert E., and Merle Y. Weiss. 1974. "The Structure of Ocean Shipping Charges," *Explorations in Economic Research*, National Bureau of Economic Research, vol. 1, no. 1 (Summer) pp. 162-193.

Appendix D

STUMPAGE PRICES

CONVERSION FACTORS USED IN CALCULATING
SAWTIMBER STUMPAGE PRICES

The value of a sawlog is dependent upon both its lumber and pulp wood values. In order to estimate stumpage values,[1] it was necessary to convert cubic meters of sawlogs into board-feet of lumber volume of pulpwood. To accomplish this the following conversion factors[2] were used:

Log-lumber conversion coefficient
(MBF lumber/cubic meters log)

12" to 17" logs	.25992
6" to 11" logs	.22001

Usable residual from lumber processing
(Percent of log volume usable as pulpwood)

12" to 17" logs	28 percent
6" to 11" logs	39 percent

Log-Pulp conversion coefficient
(Cubic meters logs/metric tons pulp)

Softwood	4.7 m^3 logs/MT pulp
Hardwood	4.1 m^3 logs/MT pulp

---

[1] The stumpage prices used throughout this study are not those found in a local market at a particular time. Rather, they are region wide "implicit" prices based upon world markets estimated by the process outlined below.

[2] Calculated from Hartman and coauthors, <u>Conversion Factors for the Pacific Northwest Forest Industry</u>, Institution of Forest Products, College of Forest Resources, University of Washington, Seattle, Washington, no date.

Throughout the study the Pacific Northwest (PNW) was assumed to have the larger (12" to 17") logs, while all the other regions were assumed to have the smaller (6" to 11") logs. This is in keeping with the expectation that older PNW logs will be substantially larger.

STUMPAGE PRICE CALCULATION PROCEDURE AND STUMPAGE PRICE TABLES FOR BASE CASE AND SENSITIVITY ANALYSIS SCENARIOS

Calculation Procedure

The procedure for calculating base pulpwood and sawtimber stumpage prices consists of adjusting the market price of the final product by subtracting from it the market prices of all the various production inputs until a residual, representing the stumpage price, is calculated. These costs are international transport, pulp mill and sawmill processing costs, and harvest and local transport costs.

The following is an illustration of this procedure using Loblolly pine (<u>Pinus taeda</u>) for the U.S. South.

Calculation of Base Stumpage Prices for Loblolly Pine in the U.S. South

For the calculation presented below, the relevant goods are lumber and bleached kraft pulp, and the relevant market for the U.S. South is the northeastern United States.

Market price of lumber, CIF New York	$306.4/MBF (1,000 bd-ft)
Transportation from mill to New York	-28.8
Mill value of lumber	$277.6/MBF
Mill processing costs	133.0
Value of wood content of lumber	$144.6

A cubic meter (m^3) of 6-in. to 11-in. sawlogs is transformed into 220.01 board-feet of lumber plus residuals that can be used as raw material for the pulping process. That part of the cubic meter that is sawn into lumber thus has a value of:

$$\frac{.22001 \text{ MBF lumber}}{m^3 \text{ sawlog}} \times \frac{\$144.6}{\text{MBF lumber}} = \frac{\$31.81}{m^3 \text{ sawlog}}$$

The value of a cubic meter of raw wood used for pulping is similarly derived:

APPENDIXES

Market price of pulp, CIF New York	$438.4/MT (metric ton)
Transportation from mill to New York	-18.0
Mill value of pulp	$420.4/MT
Mill processing costs	275.0
Value of wood content of pulp	$145.4

Since 4.7 cubic meters of raw wood are needed to manufacture 1 metric ton of pulp, a cubic meter of wood entering the pulp process has a value of:

$$\frac{\$145.4}{\text{MT pulp}} \div \frac{4.7 \text{ m}^3 \text{ Pulpwood}}{\text{MT Pulp}} = \frac{\$30.94}{\text{m}^3 \text{ pulpwood}}$$

Since 39 percent of the cubic volume of 6-in. to 11-in. sawlog is a pulpable residual of the sawmilling process, the value of the residual is:

$$.39 \times \$30.94 = \$12.07$$

The delivered price of sawlogs to an integrated sawmill/pulp mill facility is thus:

$$\$31.81 + \$12.07 = \$43.88/\text{m}^3$$

The delivered price of pulpwood is as above, $30.94/m^3.

The stumpage prices for pulpwood and sawtimber can be found by subtracting harvest and woods-to-mill transportation costs from the delivered wood prices. These costs are higher for wood resulting from a thinning operation than for wood that has been clear-cut because of smaller piece sizes and scale diseconomies.

Thus, for wood coming from a thinning operation the stumpage prices are:

Delivered price	$43.88
Harvest and transport	16.00
	$27.88/m^3 sawtimber stumpage
Delivered price	$30.94
Harvest and transport	16.00
	$14.94/m^3 pulpwood stumpage

For a clear-cut operation, the stumpage prices are:

Delivered price	$43.88
Harvest and transport	12.00
	$31.88/m^3 sawtimber stumpage

```
    Delivered price              $30.94
    Harvest and transport         12.00
                                 $18.94/m³ pulpwood stumpage
```

STUMPAGE PRICES FOR BASE CASE AND SENSITIVITY ANALYSIS SCENARIOS

The following tables present the stumpage prices (in 1979 U.S. dollars per cubic meter) that were used in the base case and sensitivity analysis scenarios. The prices vary depending on whether they are from a final harvest or commercial-thin (higher commercial-thin harvest costs depress stumpage prices) and on whether the stumpage is pulpwood or sawtimber. For the nontraditional producers the "second-quality sawtimber" price for long-fiber species is discounted 10 percent.

Table D-1. Stumpage Prices—Base Case
($ per m³)

Region/species	Commercial Thin			Harvest		
	Pulpwood	Sawtimber	2nd-quality sawtimber	Pulpwood	Sawtimber	2nd-quality sawtimber
North America						
U.S. South						
Pinus taeda, avg.-yield site	14.94	27.88	n.a.	18.94	31.88	n.a.
Pinus taeda, high-yield site	14.94	27.88	n.a.	18.94	31.88	n.a.
Pacific Northwest						
Pseudotsuga menziesii, avg.-yield site	n.a.	n.a.	n.a.	14.26	23.79	n.a.
Pseudotsuga menziesii, high-yield site	10.26	19.79	n.a.	14.26	23.79	n.a.
South America						
Brazil, Amazonia						
Pinus caribaea	14.26	26.49	19.34	18.26	30.49	23.34
Gmelina spp.	7.71	n.a.	n.a.	11.71	20.79	n.a.
Brazil, Central						
Eucalyptus spp.	7.71	16.79	n.a.	11.71	20.79	n.a.
Brazil, Southern						
Pinus taeda	14.26	26.49	19.34	18.26	30.49	23.34
Chile						
Pinus radiata	11.45	20.75	13.60	15.45	24.75	17.60
Oceania						
Australia						
Pinus radiata	9.66	17.09	10.58	13.66	21.09	14.58
New Zealand						
Pinus radiata	9.66	17.09	10.58	13.66	21.09	14.58
Africa						
South Africa						
Pinus patula	12.55	23.01	15.86	16.55	27.01	19.86
Gambia-Senegal						
Gmelina spp.	8.29	17.87	n.a.	12.29	21.87	n.a.
Eucalyptus spp.	8.29	n.a.	n.a.	12.29	21.87	n.a.
Europe						
Nordic						
Picea abies	16.98	32.06	n.a.	20.98	36.06	n.a.
Asia						
Borneo						
Pinus caribaea	11.53	20.92	14.42	15.53	24.92	18.42

Table D-2. Stumpage Prices for Sensitivity Analyses—Low Harvest Costs
($ per m³)

Region/species	Commercial Thin			Harvest		
	Pulpwood	Sawtimber	2nd-quality sawtimber	Pulpwood	Sawtimber	2nd-quality sawtimber
North America						
U.S. South						
Pinus taeda, avg.-yield site	18.14	31.08	n.a.	21.34	34.08	n.a.
Pinus taeda, high-yield site	18.14	31.08	n.a.	21.34	34.08	n.a.
Pacific Northwest						
Pseudotsuga menziesii, avg.-yield site	n.a.	n.a.	n.a.	16.66	26.19	n.a.
Pseudotsuga menziesii, high-yield site	13.46	22.99	n.a.	16.66	26.19	n.a.
South America						
Brazil, Amazonia						
Pinus caribaea	17.46	26.69	22.54	20.66	32.89	25.74
Gmelina spp.	10.91	n.a.	n.a.	14.11	23.19	n.a.
Brazil, Central						
Eucalyptus spp.	10.91	19.99	n.a.	14.11	23.19	n.a.
Brazil, Southern						
Pinus taeda	17.46	29.69	22.54	20.66	32.89	25.74
Chile						
Pinus radiata	14.65	23.96	16.80	17.85	27.15	20.00
Oceania						
Australia						
Pinus radiata	12.86	20.29	13.78	16.06	23.49	16.98
New Zealand						
Pinus radiata	12.86	20.29	13.78	16.06	23.49	16.98
Africa						
South Africa						
Pinus patula	15.75	26.21	19.06	18.95	29.41	22.26
Gambia-Senegal						
Gmelina spp.	11.49	21.07	n.a.	14.69	24.27	n.a.
Eucalyptus spp.	11.49	n.a.	n.a.	14.69	24.27	n.a.
Europe						
Nordic						
Picea abies	20.18	35.26	n.a.	23.38	38.46	n.a.
Asia						
Borneo						
Pinus caribaea	14.73	24.12	17.62	17.93	27.32	20.82

Table D-3. Stumpage Prices for Sensitivity Analyses—High Harvest Costs
($ per m³)

Region/species	Commercial Thin			Harvest		
	Pulpwood	Sawtimber	2nd-quality sawtimber	Pulpwood	Sawtimber	2nd-quality sawtimber
North America						
U.S. South						
Pinus taeda, avg.-yield site	11.74	24.68	n.a.	16.54	29.48	n.a.
Pinus taeda, high-yield site	11.74	24.68	n.a.	16.54	29.48	n.a
Pacific Northwest						
Pseudotsuga menziesii, avg.-yield site	n.a.	n.a.	n.a.	11.86	21.39	n.a.
Pseudotsuga menziesii, high-yield site	7.06	16.59	n.a.	11.86	21.39	n.a.
South America						
Brazil, Amazonia						
Pinus caribaea	11.06	23.29	16.14	15.86	28.09	20.94
Gmelina spp.	4.51	n.a.	n.a.	9.31	18.39	n.a.
Brazil, Central						
Eucalyptus spp.	4.51	13.59	n.a.	9.31	18.39	n.a.
Brazil, Southern						
Pinus taeda	11.06	23.29	16.14	15.86	28.09	20.94
Chile						
Pinus radiata	8.25	17.55	10.40	13.05	22.35	15.20
Oceania						
Australia						
Pinus radiata	6.46	13.89	7.38	11.26	18.69	12.18
New Zealand						
Pinus radiata	6.46	13.89	7.38	11.26	18.69	12.18
Africa						
South Africa						
Pinus patula	9.35	19.81	12.66	14.15	24.61	17.46
Gambia-Senegal						
Gmelina spp.	5.09	14.67	n.a.	9.89	19.47	n.a.
Eucalyptus spp.	5.09	n.a.	n.a.	9.89	19.47	n.a.
Europe						
Nordic						
Picea abies	13.78	28.86	n.a.	18.58	33.66	n.a.
Asia						
Borneo						
Pinus caribaea	8.33	17.72	11.22	13.13	22.52	16.02

Table D-4. Stumpage Prices for Sensitivity Analyses—High Transportation Costs (World Market Prices Unchanged) ($ per m³)

Region/species	Commercial Thin			Harvest		
	Pulpwood	Sawtimber	2nd-quality sawtimber	Pulpwood	Sawtimber	2nd-quality sawtimber
North America						
U.S. South						
Pinus taeda, avg.-yield site	14.17	26.31	n.a.	18.17	30.31	n.a.
Pinus taeda, high-yield site	14.17	26.31	n.a.	18.17	30.31	n.a.
Pacific Northwest						
Pseudotsuga menziesii, avg.-yield site	n.a.	n.a.	n.a.	12.06	18.88	n.a.
Pseudotsuga menziesii, high-yield site	8.06	14.88	n.a.	12.06	18.88	n.a.
South America						
Brazil, Amazonia						
Pinus caribaea	12.86	23.64	16.04	16.86	27.64	20.04
Gmelina spp.	6.10	n.a.	n.a.	10.10	18.03	n.a.
Brazil, Central						
Eucalyptus spp.	6.10	14.03	n.a.	10.10	18.03	n.a.
Brazil, Southern						
Pinus taeda	12.86	23.64	16.04	16.86	27.64	20.04
Chile						
Pinus radiata	9.49	16.75	7.12	13.49	20.75	11.12
Oceania						
Australia						
Pinus radiata	8.13	13.87	6.48	12.13	17.87	10.48
New Zealand						
Pinus radiata	8.13	13.87	6.48	12.13	17.87	10.48
Africa						
South Africa						
Pinus patula	10.81	19.46	11.21	14.81	23.46	15.21
Gambia-Senegal						
Gmelina spp.	6.80	15.15	n.a.	10.80	19.15	n.a.
Eucalyptus spp.	6.80	n.a.	n.a.	10.80	19.15	n.a.
Europe						
Nordic						
Picea abies	16.13	30.32	n.a.	20.13	34.32	n.a.
Asia						
Borneo						
Pinus caribaea	10.00	18.41	11.50	14.00	22.41	15.50

Table D-5. Stumpage prices for Sensitivity Analyses--High Transportation Costs (World Market Prices Adjusted) ($ per m³)

Region/species	Commercial Thin			Harvest		
	Pulpwood	Sawtimber	2nd-quality sawtimber	Pulpwood	Sawtimber	2nd-quality sawtimber
North America						
U.S. South						
Pinus taeda, avg.-yield site	14.78	26.81	n.a.	18.78	30.81	n.a.
Pinus taeda, high-yield site	14.78	26.81	n.a.	18.78	30.81	n.a.
Pacific Northwest						
Pseudotsuga menziesii, avg.-yield site	n.a.	n.a.	n.a.	13.16	20.51	n.a.
Pseudotsuga menziesii, high-yield site	9.16	16.51	n.a.	13.16	20.51	n.a.
South America						
Brazil, Amazonia						
Pinus caribaea	13.94	25.18	17.89	17.94	29.18	21.89
Gmelina spp.	6.95	n.a.	n.a.	10.95	19.31	n.a.
Brazil, Central						
Eucalyptus spp.	6.95	15.31	n.a.	10.95	19.31	n.a.
Brazil, Southern						
Pinus taeda	13.94	25.18	17.89	17.94	29.18	21.89
Chile						
Pinus radiata	9.81	17.99	10.59	13.81	21.99	14.59
Oceania						
Australia						
Pinus radiata	8.45	14.42	7.14	12.45	18.42	11.14
New Zealand						
Pinus radiata	8.45	14.42	7.14	12.45	18.42	11.14
Africa						
South Africa						
Pinus patula	12.07	21.09	13.45	16.07	25.09	17.45
Gambia-Senegal						
Gmelina spp.	8.04	16.76	n.a.	12.04	20.76	n.a.
Eucalyptus spp.	8.04	n.a.	n.a.	12.04	20.76	n.a.
Europe						
Nordic						
Picea abies	17.21	31.86	n.a.	21.21	35.86	n.a.
Asia						
Borneo						
Pinus caribaea	10.32	18.86	11.98	14.32	22.86	15.98

Table D-6. Stumpage Prices for Sensitivity Analyses--High Lumber-Processing Costs ($ per m³)

Region/species	Commercial Thin			Harvest		
	Pulpwood	Sawtimber	2nd-quality sawtimber	Pulpwood	Sawtimber	2nd-quality sawtimber
North America						
U.S. South						
Pinus taeda, avg.-yield site	14.94	17.46	n.a.	18.94	21.46	n.a.
Pinus taeda, high-yield site	14.94	17.46	n.a.	18.94	21.46	n.a.
Pacific Northwest						
Pseudotsuga menziesii, avg.-yield site	n.a.	n.a.	n.a.	14.26	13.60	n.a.
Pseudotsuga menziesii, high-yield site	10.26	9.60	n.a.	14.26	13.60	n.a.
South America						
Brazil, Amazonia						
Pinus caribaea	14.26	16.07	8.92	18.26	20.07	12.92
Gmelina spp.	7.71	n.a.	n.a.	11.71	9.70	n.a.
Brazil, Central						
Eucalyptus spp.	7.71	5.70	n.a.	11.71	9.70	n.a.
Brazil, Southern						
Pinus taeda	14.26	16.07	8.92	18.26	20.07	12.92
Chile						
Pinus radiata	11.45	10.34	3.19	15.45	14.34	7.19
Oceania						
Australia						
Pinus radiata	9.66	6.67	0.17	13.66	10.67	4.17
New Zealand						
Pinus radiata	9.66	6.67	0.17	13.66	10.67	4.17
Africa						
South Africa						
Pinus patula	12.55	12.59	5.44	16.55	16.59	9.44
Gambia-Senegal						
Gmelina spp.	8.29	6.79	n.a.	12.29	10.79	n.a.
Eucalyptus spp.	8.29	n.a.	n.a.	12.29	10.79	n.a.
Europe						
Nordic						
Picea abies	16.98	21.65	n.a.	20.98	25.65	n.a.
Asia						
Borneo						
Pinus caribaea	11.53	10.50	4.00	15.53	14.50	8.00

Table D-7. Stumpage Prices for Sensitivity Analyses—High Pulp—Processing Costs
($ per m^3)

Region/species	Commercial Thin			Harvest		
	Pulpwood	Sawtimber	2nd-quality sawtimber	Pulpwood	Sawtimber	2nd-quality sawtimber
North America						
U.S. South						
Pinus taeda, avg.-yield site	3.23	23.31	n.a.	7.23	27.31	n.a.
Pinus taeda, high-yield site	3.23	23.31	n.a.	7.23	27.31	n.a.
Pacific Northwest						
Pseudotsuga menziesii, avg.-yield site	-1.45	16.51	n.a.	2.55	20.51	n.a.
Pseudotsuga menziesii, high-yield site	-1.45	16.51	n.a.	2.55	20.51	n.a.
South America						
Brazil, Amazonia						
Pinus caribaea	2.55	21.92	14.77	6.55	25.92	18.77
Gmelina spp.	-5.71	n.a.	n.a.	-1.71	15.55	n.a.
Brazil, Central						
Eucalyptus spp.	-5.71	11.55	n.a.	-1.71	15.55	n.a.
Brazil, Southern						
Pinus taeda	2.55	21.92	14.77	6.55	25.92	18.77
Chile						
Pinus radiata	-0.26	16.19	9.04	3.74	20.19	13.04
Oceania						
Australia						
Pinus radiata	-2.04	12.50	6.02	1.96	16.50	10.02
New Zealand						
Pinus radiata	-2.04	12.50	6.02	1.96	16.50	10.02
Africa						
South Africa						
Pinus patula	0.85	18.45	11.30	4.85	22.45	15.30
Gambia-Senegal						
Gmelina spp.	-5.12	12.64	n.a.	-1.12	16.64	n.a.
Eucalyptus spp.	-5.12	n.a.	n.a.	-1.12	16.64	n.a.
Europe						
Nordic						
Picea abies	5.28	27.45	n.a.	9.28	31.45	n.a.
Asia						
Borneo						
Pinus caribaea	-0.17	16.36	9.86	3.83	20.36	13.86

Index

Absolute advantage, and comparative advantage, 32–33, 81
Afforestation
  in Brazil, 33
  in Europe, 6
  land for, 85
  of savanna, 86, 88
Africa
  plantation forests in, 8
  representative plantations, 41, 121–126 (table)
Agriculture, slash-and-burn, 85–86
Asia, representative plantations, 41, 129–130 (table)
Australia, plantation forests in, 7

Base case
  findings, 36–41
  with 5 percent discount rate, 37 (table)
  internal rate of return, 40 (table)
  interpretation of, 41–42
  summary of, 42 (table)
  with 10 percent discount rate, 38 (table)
Biological potential, regional, 12
Biological risks, 30, 88–90
Biological yields, 43
Brazil
  chemical pulp from, 2
  plantation afforestation in, 33
  reforestation in, 7
British Columbia. *See* Pacific Northwest

Chile, 32
  plantation forests in, 7
Comparative advantage, and absolute advantage, 32–33, 81
Costs, model versus actual, 49–50

Data sources, 14–15
Deforestation, 84
Development costs
  and present net value, 43
  by region, 29
Discount rates, 17

Economic rents, 17, 27
Europe
  representative plantations, 41, 127–128 (table)
  as wood deficit region, 18
Exchange rates, 34–35
Exotic species
  monoculture of, 30
  and pests, 89
Externalities, 31

Fertilization, 30
Forest economy, 3
Forest-producing regions, 5–8
Forest products
  demand and supply of, 3
  prices, 2, 16
Forest resources
  regional shifts, 82
  second-growth, 3–4
Forestry plantation. *See* Plantation forestry
Forest soils, 87
Fuelwood, 84, 86

Governmental policy, 21

Harvest cost
  model component, 22–23
  sensitivity analysis, 51
  sensitivity of PNV, 58 (table), 59 (table), 60 (table)
High-harvest-cost scenario
  with 5 percent discount rate, 55 (table)
  internal rate of return, 57 (table)
  with 10 percent discount rate, 56 (table)
High-lumber-production-cost scenario
  with 5 percent discount rate, 70 (table)
  internal rate of return, 72 (table)
  with 10 percent discount rate, 71 (table)
High-pulp-processing-cost scenario
  with 5 percent discount rate, 67 (table)
  internal rate or return, 69 (table)
  with 10 percent discount rate, 68 (table)
High-transportation-cost scenario
  with 5 percent discount rate, 61 (table), 64 (table)
  internal rate of return, 63 (table), 66 (table)
  with 10 percent discount rate, 62 (table), 65 (table)

Industrial forest plantation. *See* Plantation forestry
Integrated forest company, 9
Internal rates of return, 18
Internal transport cost, 22–23
  sensitivity analysis, 51
International transport cost, 20
  for lumber, 142 (table C-3)
  methodology, 138–145
  model component, 20–21
  for pulp, 141 (table C-2)
  sensitivity analysis, 51–59
Investment criteria, 78

Japan, 18
Jari project, 88
  costs, 21, 27–28

Land
  abandoned agricultural, 86
  acquisition and development costs, 26–29
  appropriate use of, 85–86
  market price of, 32
Location, choice of, 9–10
Low-harvest-cost scenario
  with 5 percent discount rate, 52 (table)
  internal rate of return, 54 (table)
  with 10 percent discount rate, 53 (table)
Lumber, 24–25
  world market prices, 56 (table)

Management regimes, 12–14
  choice of, 10–11
  in North America, 13
Model
  basic components of, 17–25
  Faustmann-type, 17
  mathematical formulation of, 134–137
Monoculture, 88
Mycorrhizal fungi, 87

New Zealand, 32
  plantation forests in, 7
North America, representative plantations, 39–40, 99–106 (table)
Northern Hemisphere, lumber stumpage prices, 78

Oceania, representative plantations, 41, 117–120 (table)
Outputs
  choice of, 10, 23–25
  mix of, 12–14

Pacific Northwest
  old-growth inventories, 16
  as surplus region, 18
  woodchip exports, 33
  as wood-exporting region, 6
Panama Canal, 86
Plantation forestry, 28–29, 44 (table)
  biological yields, 45 (table)
  decision variables, 9–11

INDEX

economic viability of, 4
environmental benefits of, 31
exotic species in, 30
and indigenous forests, 90–91
land area, 5 (table), 7 (table)
land for, 2–3
long-term supply potential, 81–82
and multiproduct firm, 10
in Northern Hemisphere, 1
pests and diseases, 88–90
regional differences in, 39–41
representative, 11–14
rotation lengths, 45 (table)
single-species, 87
site degradation, 87–88
site rehabilitation, 86–87
time frame, 10
for world markets, 3
PNV. *See* Present net value
PNW/BC. *See* Pacific Northwest
Political risks, 31
Present net value (PNV), 17–18
  economic interpretation of, 42
  and land price, 27
  sensitivity of, 48, 79 (table)
  shifts summary, 73–76
Prices
  European, 20
  free-on-board (FOB) mill, 21
Processing
  alternative structures, 33–34
  cost sensitivity analysis, 59–68
  model component, 21–22
Production functions, 12

Real-stumpage-price growth case
  with 5 percent discount rate, 73 (table), 76 (table)
  internal rate of return, 75 (table), 78 (table)
  with 10 percent discount rate, 74 (table), 77 (table)
Reforestation, 1–2, 7
  land for, 85
  in United States, 6
Representative plantations
  costs and yields, 97–133
  economic performance of, 46
Risk. *See* Biological risk; Political risk
Rotation lengths, 43

Sawnwood. *See* Lumber
Sensitivity analysis
  implications of, 77
  reasons for, 49
  summary of, 80–81
Silvicultural practices, 10–11
South America
  plantation forestry potential, 7
  representative plantations, 41, 107–116 (table)
Southeast Asia, plantation forests in, 7–8
Southern Hemisphere
  plantation forestry success in, 17
  and softwood lumber, 39
Species
  choice of, 11, 90
  extinction of, 85
Stumpage prices, 20
  base case, 151 (table)
  increase in, 68–73
  of lumber, 24
  methodology, 147–157
  as model component, 23
  and output mix, 11
  relative, 32–33
  for sensitivity analysis, 152–157 (table)

Trade. *See* World trade
Tropical forest
  growth rates, 2
  preservation of, 91

United States, forestry implications in, 82

Watershed management, 86–87
Wood, deficit regions, 3, 18–19
Wood pulp, 24
  processing costs, 28
  world market prices, 56 (table)
World trade
  duties, 23
  as model component, 18–20
  structure of, 82–83
  tariff policy, 21